John W Nystrom

On Technological Education and the Construction of Ships and

Screw Propellers

For naval and marine engineers

John W Nystrom

On Technological Education and the Construction of Ships and Screw Propellers
For naval and marine engineers

ISBN/EAN: 9783337418687

Printed in Europe, USA, Canada, Australia, Japan

Cover: Foto ©berggeist007 / pixelio.de

More available books at **www.hansebooks.com**

ON

TECHNOLOGICAL EDUCATION

AND THE

CONSTRUCTION

OF

SHIPS AND SCREW PROPELLERS,

FOR

NAVAL AND MARINE ENGINEERS.

BY

JOHN W. NYSTROM,

LATE ACTING CHIEF ENGINEER UNITED STATES NAVY.

SECOND EDITION REVISED, WITH ADDITIONAL MATTER.

PHILADELPHIA:

HENRY CAREY BAIRD,

INDUSTRIAL PUBLISHER,

406 Walnut Street.

1866.

PHILADELPHIA:
COLLINS, PRINTER, 705 JAYNE STREET.

PREFACE TO THE SECOND EDITION.

A SECOND edition of this work has been called for, and as Congress has taken no action on the subject, the author avails himself of the opportunity to add further arguments in favor of technological education. The topic cannot be too much discussed, and certainly merits the serious consideration of every good citizen.

There are many engineers in the Navy who would be equally disposed to agitate this subject and assist in the reorganization and improvement of the corps, but their position restrains them, and they cannot expose themselves to the ungrateful task, which is unavoidable in the elucidation of the existing system. The lot seems to have fallen on the author to take the bull by the horns, and he fearlessly courts an open contest, if such can only be had, with the organized prejudices which now embarrass the subject. If in fault, he is open for correction.

1*

It is to be regretted that this theme cannot be fairly treated without attacking, as it were, functionaries who only happen to be illustrations of a decrepit system; but it is to be hoped that those gentlemen who have been thus particularized will join in the proposition at issue.

If the engineers themselves are not sufficiently alive to the importance of this subject from a deficiency of experience in applied science, of course it will not be expected that politicians, who have no light to guide them but a sense of duty, can possibly assume the initiation of any such reformations as these.

What is true in reference to the interests of the Federal Government, applies with equal force to the civil interests of all the separate States, each of which ought to have its Technological Institute, to give the entire profession of Civil and Mechanical Engineers that completeness of qualification for their career, which by the existing system is so rarely and so imperfectly attained.

PHILADELPHIA, March 28, 1866.

PREFACE TO THE FIRST EDITION.

MUCH consideration has been given to the propriety of publishing the accompanying views on technological education, as they were originally not intended for that purpose; but as some steps must be taken in that direction before long, in compliance with what many deem an imperious public necessity, the hesitation was at length relinquished. In order to render the necessity of technological education more conspicuous, occasional reference has been made to actual cases of engineering disappointments and mismanagement, growing out of a want of applied science. Engineers are often intrusted with responsible stations, without being possessed of adequate knowledge of their profession, or without having gradually and fairly earned such appointment in the field of experience. It is not yet time to attempt to classify the rank and position of engineers in the several departments, civil and military, on land and sea, as, for obvious reasons, it must be deferred.

The relation between engineers and sailing officers on board of steamers has, in all countries, been a troublesome question, ever since the introduction of steam. The engineer knows himself to be in a very responsible position, not always appreciated by his captain. He is often of very limited education, and when he finds himself imposed upon, perhaps inadvertently uses stronger language than necessary in his defence, which has often been the cause of discord.

Education is necessary to the engineer, not, however, principally for enabling him to please the captain, but for the proper performance of his professional duty generally, and he will, at the same time, accord and harmonize better with the sailing officers by whom he is surrounded.

The Corps of Engineers in the United States

Navy is on a better footing than that of any one in Europe, but, nevertheless, it does not enjoy the standing due to its important office, either in respect to its rank or its education. The United States Navy has now taken the lead in the new tactics of naval warfare, and through a decisive experience has developed the necessity of making a corresponding change in its executive organization, as regards the rank and learning of its engineers.

What is required here cannot be learned in foreign countries, for whilst our present experience is far ahead of theirs, neither their practical knowledge nor their accepted theories are sufficient or altogether applicable to our case. It is therefore necessary for the Corps of Engineers, relying only upon their characteristic enterprise and independence of mind, to carry their achievements still further onwards, and by qualifying themselves to maintain with dignity their appropriate rank in the service, at the same time preserve their well-earned position as pioneers in their professional career.

The writer has for many years felt the

greatest interest in the progress and standing of this Corps, and respectfully begs leave to submit herewith, for their consideration, some views on this subject, parts of which have already been communicated to the Congressional Committees on Naval Affairs, as also a further communication to the Secretary of the Navy.

JOHN W. NYSTROM.

CONTENTS.

2

CHAIRMAN OF THE COMMITTEE

OF

NAVAL AFFAIRS, U. S. CONGRESS,

WASHINGTON, D. C.

SIR: The object of this paper is to invite the attention of your Committee, and of Congress, to a subject of general interest to the country, and one of particularly great importance to the power and prosperity of its navy.

The subject is that of establishing a *Technological Academy for Naval Engineers*, and for the *promotion of sciences* bearing on the immediate necessities of the country in that Department. Should it receive your Committee's attention and approbation, and should you consider it worthy of being submitted to Congress, the undersigned is willing to suggest the requisite plans and method for its organization.

I have the honor to remain,

Your obedient servant,

JOHN W. NYSTROM,

Engineer.

MARKOE HOUSE,
PHILADELPHIA, Dec. 21, 1863.

TECHNOLOGICAL EDUCATION.

THE immense natural resources of the New
World are confided to the hands of an enterprising, ingenious, and happy people; yet their
time, their money, their life, and their credit,
in imitation of the Old World, are lavishly
wasted, through a deficient knowledge of those
physical laws which constitute the most essential element of all human enterprise. Under
this impression the writer has striven, by means
of various scientific articles on these subjects,
to enforce the necessity of enlarged reform,
both in the study and the application of these
laws, which he modestly hopes may have some
good effect.

The efforts of a single individual, however,
when elevated to subjects of such magnitude,
only result in perpetual and unprofitable
struggles with organized interests and prejudices, and fail of their purpose through the

2

misconceptions which are inseparable from new
and original subjects. From its novelty alone,
a new and valuable idea is frequently condemned on bare supposition, and the writer has
thus labored in vain, under the greatest disad-
vantages (accompanied with great expense), to
rescue the proposition which he is now about
to submit to your committee, from that neglect
to which, for many years, it has been doomed
by the indifference or imperfect appreciation
of those around him.

Up to the present day, the knowledge of
steam engineering, in which we take so much
pride, and which constitutes a most essential
part of our national existence, is far behind our
general knowledge of science. Our marine
engines and boilers are not only unnecessarily
complicated, but prodigally extravagant in
their consumption of fuel; whilst not unfre-
quently new machinery fails to accomplish expected results from the want of knowledge of
the physical laws bearing on the problem.

Only a few years ago there was not a single
"steam propeller" in the United States with
properly constructed air-pumps and foot-valves.
Some of the propellers designed to go to Europe succeeded in making one passage, whilst
others broke down at but a short distance from

the shore, and returned; most of them existed
but a short time, and involved the loss of millions of dollars to the country, to say nothing
of the effect upon its scientific and mechanical
reputation. Enterprising merchants, who attempted to establish lines of steamers to Europe,
became discouraged, and perhaps ruined by
their failure, and the result now is that we
have not a single steamer in the European
trade. A few names of steamers may be re-
ferred to in verification of these remarks—
namely, the frigate *San Jacinto*, which broke
down through disarrangement of her air-pumps
and foot-valves; the *La Fayette* also (whose
machinery contained, perhaps, the worst air-
pumps ever constructed); the *City of Petersburg;*
Ben Franklin; and the *Frigate Merrimac*, which
suffered from the breaking down of her foot-
valves on her passage from America to England, in 1856, and which had to be improved
in England. In addition to these, a great many
other first-class American steamers experienced
the same fate for want of *applied physical science.*

When the writer became acquainted with
these defects, he attempted to correct them by
explaining the physical laws in operation,
which, at the time, was only received with
ridicule, and derided as theoretical. He then

worked out practical 'formulas, which were published in the Journal of the Franklin Institute, of Philadelphia, and in his "Pocket-Book of Engineering," giving a solution of the principles under which the air pump and foot-valves operate, after which many errors were corrected; but up to the present day there is no other publication on the subject. That publication has been copied and republished in Europe.

Blunders of this kind are still going on, by means of which millions of dollars are wasted, and our national reputation impaired from this general want of applied *physical science*.

The physical laws connected with the operation of fresh water or surface condensers, with the combustion of fuel, with the nature and properties of steam; with the dynamic equivalent, the conducting economy, and other properties of heat, are yet but partly known, and that by but a few scientific men in the world; and in no case are they worked out to a practical shape, with formulas intelligible to the engineers in the shop.

We find by science that the theoretical effect of *one pound* of pure carbon consumed per hour, is over *five horses*, whilst in our present practice it requires several pounds of coal per hour for each horse power.

Fresh water condensers have always given trouble; the combustion of fuel requires most earnest attention, not so much for the cost of fuel, as to enable us to navigate long distances with great speed, and with something more on board than boilers, machinery, and fuel. The Navy Department is now having steamers built intended for great speed, the arrangements of which plainly show a want of proper knowledge in steamship performance.

The knowledge of steamship performance is yet at a very low point, and for want of it no accurate record can be kept, by which to compare the true quality of performance of one steamer with that of another; or to determine what will be the performance of a steamer constructed according to given data.

Ingenious contrivances in machinery and steamers, with plausible promises of high speed (up to twenty and thirty miles per hour), are frequently met with, whose plans, sometimes confidently accepted, often result in complete failure and disappointment, which properly applied science would have avoided. The truth of this observation will probably be realized by a wild scheme now before the City Councils of Philadelphia, proposing a line of steamers to Europe, which has for several years remained

in a nebulous condition, unsupported by that
scientific reasoning by which alone any one
could be rendered confident of the result.

Some twelve years ago, the writer was com-
piling a Pocket-Book of Mechanics and Engineering (since published in repeated editions);
he analyzed many previously published data,
and, with the aid of his own experience, reduced the law of steamship performance to a
practical rule to work by.   In one of the most
respectable journals of the country were found
some plausible data on steamship performance,
which threw the writer into the utmost confusion, and in their solution involved him in
great expense of travel and practical investigation, only to find them bold exaggerations.

The Navy Department have made very extensive experiments on the expansion of steam,
which were commenced in New York some
four years ago.   The well-known Erie expansion experiments, the Washington Navy Yard
and Old Point boat experiments, and the expansion experiments made lately at the Novelty
Iron Works, N. Y., were all carried on in the
apparent attempt to overthrow natural laws,
and establish physical by-laws. The engineers
are manifestly not familiar with the scientific
principles which belong to the question.

The natural effect in a given quantity of steam, of given temperature and pressure, is as specific as the natural effect in a waterfall. We have only to strive, by improvements in the arrangement of our steam-engines and boilers, to utilize the greatest possible percentage of that natural effect, as is done by water-wheels and turbines. If the Navy Department find no utility in the expansion of steam, it only exposes its position in the science of steam engineering, which can be no indication of what may be done by other parties.

Steam-engines, like water-wheels and turbines, utilize widely different percentages of the natural effect, even with equal grade of expansion, which may be seen in the results of the different experiments made on different engines by the Navy Department. The result of each experiment is applicable only to that peculiar arrangement of the engine and boiler experimented upon, and no more.

The maximum or natural work K, per unit of heat in steam, is in footpounds.

$$K = \frac{144\ P\ (V-1)\ (2 \cdot 3\ log. \dfrac{S}{l} + 1)^*}{H'\ V}.$$

* See Nystrom's Pocket-Book, 10th edition, for the value of these quantities.

P = total steam pressure per square inch.

H' = units of heat per cubic foot of the steam P.

V = volume of the steam compared with water.

S = stroke of steam-piston, in inches.

l = part of the stroke under which steam is fully admitted, in inches.

The natural effect of the heat in the steam in horse-power, will be—

$$= \frac{N K}{550\ T}$$

N = total number of units of heat passed through the steam-engine in the time T in seconds. The more of this natural effect that can be utilized, the more perfect is the steam-engine. It is time to speak about steam-engines as we do about water-wheels and turbines— namely, "how many per cent. it utilizes of the natural effect."

On the writer's last arrival from Europe, Dec. 1860, he found the anti-expansion question receiving considerable attention by engineers. He published, in a scientific journal, some demonstrations to prove the folly of the Erie expansion experiments; and, although ridiculed in a New York paper, they produced good effect. Again, he published in his Pocket-Book tables for expansion of steam (hitherto the most complete in print), and its connection with superheated steam, which also had a great effect, as subsequent experiments resulted in their favor.

The Navy Department is here alluded to because their blunders are more perfectly exposed to view, and therefore better known, but the evil is none the less serious in private enterprises.

The writer is in possession of knowledge which would greatly contribute to clear up these difficulties, and advance the character of our steam engineering, but he cannot undertake the great expense of bringing it before the public, inasmuch as scientific knowledge is not sufficiently diffused among our mechanics and engineers to render such a work self-sustaining. In proof of which he would remark that publishers are not willing even to get up such expensive books as his "Pocket-Book of Mechanics and Engineering," of which copies are sent herewith.

The manuscript of this book was submitted to publishers in the year 1853, some of whom had it examined by scientific and practical men, who condemned it as useless, and unfit for publication. Some publishers objected to the great expense in bringing it out, whereupon it was carried through at the author's own expense, which has now amounted to a considerable sum. The small profit realized is not sufficient for the expense of experiments and investigation attending each succeeding edition.

I take it for granted that, in a matter so important to the profession and the country, many others, much more highly qualified by their abilities and attainments than myself, would cheerfully co-operate in the efforts to further the main purpose here foreshadowed, if an organized shape could only be given to it, or a nucleus of some kind formed upon which their efforts would be concentrated; and in view of the great expense attending it, I consider it necessary, in due regard to the interests of the engineering profession, to lay the matter before your committee, trusting that it may receive due consideration, and that, possibly, means may be appropriated for that purpose.

The labor attending the investigation of new and original subjects is immense; particularly in exploring unknown regions of science, and bringing the products home to simple formulas and tables of a practical shape. There does not, at present, appear to be any one among us who is willing or able, perhaps for want of time, to undertake such a laborious task, and very few know what is wanted, but too many suppose we have attained perfection.

We have plenty of scientific books, mostly written by professors in colleges, having very little or no opportunity to apply their knowledge in practice, and which are, therefore, destitute of practical examples.

We frequently find most valuable formulas given by scientific men in such a shape that it requires to know more than the author in order to employ them; they are not only not trimmed to a practical shape, but even the meaning of letters is rarely explained in a technical language.

It is surprising to see how successfully mathematicians have contrived to keep the simple science of the "calculus" such a perfect mystery. It reaches very few among us, not from difficulty in learning it, but simply for want of its judicious application in practice. We find books on the calculus of several hundred pages without a single practical example, which makes the science difficult and tedious of acquisition, and when acquired, very rarely further developed, but is stored away in the mind so that it cannot be found when wanted. We find simple formulas occupying several pages in explanation, which, by a solitary example applied to practice, would imprint it indelibly upon the student's memory. All this can be effectually corrected and improved by the establishment of proper institutions for the instruction of combined theory and practice.

There is now a very distinct line drawn between scientific and practical men; the more we study and cultivate the branches separately, the more distinct will this line become, and the less will they understand one another, and may ultimately fall into irreconcilable estrangement. The prejudice against science is, in our day, a very serious evil.

Science is almost despised by many practical men, not always for want of valuation of it, but often because they do not understand it. A blind man can walk on roads and streets, but when he finds an obstacle must stop; at a ditch he may tumble down into it, he cannot turn from his accustomed track. Such is the case with many practical and otherwise most valuable men working without a knowledge of physical laws. In order to follow up the improvements of the age, the track pursued by our fathers must often be abandoned, and a new one selected and surveyed for ourselves.

Without the application of science we go ahead without knowing where we are going. In verification of which we have plenty of examples in engineering blunders, sometimes subjected to a committee of inquiry, which may result in the discharge of the engineer, accompanied by extravagant abuse of the department concerned, and the evil only temporarily remedied by substituting another, who will most likely not repeat the same blunders, but will do something worse. There is yet no attempt made to permanently remove these evils and secure success in our enterprises by proper institutions. This your committee will admit to be true, but may ask " how can the evil be removed and permanently corrected ?"

America has taken the lead of the world in popular education. Its institutions are copied and imitated in Europe, but it remains for us to follow up and take the lead in the nobler and purer refinements of our nature. We have the best materials in the world by which to accomplish this object, the question is only as to the time and the means to be taken.

We are a new people; our habits and circumstances are different from those of other

3

nations, and our institutions must be organized accordingly.

In Europe they have institutions for the diffusion of combined theoretical and practical learning, the want of which is most severely felt in this country. Institutions of that kind are of more importance in America than elsewhere, for the reason that mechanical skill and inventive ingenuity are here more extensively developed, and the want of applied science wastes away a proportionate quantity of time and money.

It is very evident that we are behind some other nations in science, and, at the same time, it is certain that we have more science than we can properly manage or utilize. It is the *application of science to practice* which requires immediate attention and special institutions.

It is very gratifying to know that Congress, at its last session, passed a bill to establish a *National Academy of Science*, which will no doubt be of great value; but how can it be brought to bear advantageously on the general interest and immediate wants of the country.

Considering the peculiar circumstances in which the country is now placed (the natural fruit of time and civilization), technological institutions are absolutely necessary to enable us to rise gradually and surely to the position due to us among nations, and when once so raised, we would never fall, but become able to maintain with true dignity a *National Academy of Science*.

Technological institutions will reveal and develop the talent and ability of the nation, and bring its immense natural resources to account. It is technological institutions which alone furnish *proper materials* for a National Academy of Science. We have now among us many Newtons, Keplers, Berzelius's, Watts's, Fultons, &c. &c., but have no means of bringing them out; but, on the contrary, plenty of ingenious contrivances to screen them from observation. They are not willing to enter into competition with our everyday rivals, while our national leaders, in their most earnest exertion to find the right man for the right place, are continually imposed upon. This evil cannot be removed by the peculiar liberty alone, in which we take so much pride, but simply by a diffusion of useful knowledge through established institutions, which should constitute the true object of our national pride.

At the present time, scientific attainments and true practical knowledge are very little respected; physical laws, established by the

Creator of the universe, are often derided as theoretical; ignorance has taken the lead, and rules in the ascendant, and often adopts that which is opposite alike to science, experience, and common sense.

The object of this paper, therefore, is to propose the establishment of a *National Technological Academy* of a high order, whose purpose should be, by the combination of practical and theoretical instruction, to subserve a great public want, and at the same time to inaugurate a new era in the scientific and practical reputation of the American people.

An institution of that kind cannot be a private enterprise, for in order to command the respect necessary to its existence and high purpose, it must be a public institution.

The writer has been educated at the Royal Technological Institute, at Stockholm, where they have a complete set of workshops and laboratory, for the practical training of students between lecture hours. It is not expected, neither is it necessary that the student shall become an accomplished mechanic, but the object is to concentrate his mind on the work about which he is studying and calculating. When confined only to books and blackboards, his conceptions rarely extend any further. He acquires the knowledge by routine, as it were; the study becomes tedious to him, and when brought to bear on practice, the most simple problem may confound him. When a student is brought up in the combined science and practice, however, he generally acquires a taste for work—good workmanship and proper proportions—and the application of his science becomes a pleasure. He studies mathematics at the same time he learns drawing; physics and mechanics at the same time he makes his tools and models for machinery. His science is applied as fast as it is acquired, and he will never forget it. When a student is thus equipped for his journey of life, he is able to bring such physical laws into action as to secure success in all his enterprises. He will be able to record and report back to the institute his future experience, by which the most thorough connection may be kept up, between science and practice.

As things now stand, a man of most valuable information is not thus able to record his achievements; in fact, he may not know himself the very laws of his success; his experience and valuable knowledge die with him; his toiling successor will reiterate his blunders, and gain new experience by a new series of expensive trials and error.

3*

Steam-engineering and ship-building are arts in which we take the greatest pride; still there is no institution in the country where we learn to construct a steamer completely, or acquire the physical laws under which it operates. Ship-building and steam-engineering are yet considered different professions, while they are so intimately connected in steamboats that it would be impossible to trace a line of separation between them. The shipbuilder cannot properly construct a steamer without the knowledge of the machinery, neither can the engineer construct the machinery without a knowledge of the vessel; yet we rarely find one who can undertake both, and the result is a discord of action. They do not please one another, and neither of them takes that care in the whole arrangement which one controlling mind would do. In iron shipbuilding the two branches are more generally brought under one mind.

We rarely find a superintendent or proprietor in a machine shop or shipyard, even in our navy yards, who can master an algebraical formula, or who is in possession of the rudiments of the science bearing on his profession. We have no school where we learn to make a proper working drawing, but students are taught to color drawings before they know how to construct a shadow; the surface of everything is learned, and the substance obscured. We never find a complete working drawing of a steamer when its building is commenced! In some cases, and even in the navy yards, *the drawing is made after the steamer is finished*, when an extra bill for alterations and experiments augments the originally intended cost to an unsatisfactory sum, and often results in a complicated arrangement of machinery, with donkies, fans, pumps, cocks, and pipes placed about the vessel, here and there, like scattered stumps and logs in a forest, and requires a more skilful engineer to keep it in order than the one who contrived it. In fact, ingenuity seems to surmount any obstacle that could possibly be encountered, for, in many cases, it shows no disposition whatever to prevent or avoid the difficulty by application of proper principles at the outset; but a machine, on the contrary, is invented by which to overcome the obstacle, and the aggregate contrivance is denominated "practical."

In verification of this, we have many examples in the navy, but I will here refer to a new iron steamer, built for the merchant service, whose machinery is one of those ingenious contrivances we frequently meet with; the slide valves alone, for only one cylinder, are operated by twenty connecting rods and forty journals, occupying a height of some thirty-five feet in the vessel, and even then the engine cannot be reversed without throwing the machinery out of gear, and reversing it by hand.

We must, in all ages and in all countries, expect active and operative minds to come forward with ingenious contrivances, sometimes with wild ideas, ridiculous in design, and wrong in mechanical principles; but then it is the function of science and knowledge to step in and correct their aberration, or, if necessary, to guard against or prevent their further introduction until developed to an educated design, which otherwise might lead to destruction of life and property.

On the other hand, most ingenious and valuable ideas are sometimes submitted to the opinion of scientific men with no practical knowledge, who may condemn them from an imperfect perception of their merit. It is only a knowledge of the combined theory and practice that can accomplish justice in all cases.

The great prospect now opening before us in the present new era of naval architecture, as connected with the new national navy yard (to be established at *League Island*, we hope), will necessarily, at some time, concentrate our serious attention upon the establishment of proper institutions, and a systematic corps of naval engineers; but when will this necessity become recognized, and the proper policy be pursued? Can it be accomplished by the efforts of conciliatory reasoning, or must it be forced upon us by the suffering and losses consequent upon engineering blunders, in which our present experience does not seem sufficient to bring us to the point?

The science of war has been taught by disasters, and has been gradually advanced by the force of proper institutions, and thorough discipline to its present perfection.

The navy has lately undergone a great and very important change, and is converted into an entirely new school, by the introduction of steam and armored vessels, which change has already reduced considerably the number and length of the ropes in the old school. America has taken the lead in this new direction, and is the first nation on the globe which has brought the new naval school to the severest test, and demonstrated the necessity of a corresponding system of education. The old school is now proved to be incapable of conducting our new naval tactics.

In the army there has been no such sudden change, but the old school has been gradually improved to its present condition by the instrumentality of a properly organized corps of engineers, raised from the school-trenches to the highest accomplishment, and to the elevated rank which is due to their profession.

The works in the Departments of Ordnance, Fortification, and the Coast Survey, are of the highest order of science brought to a practical shape, unequalled in Europe.

Now let us ask, on the other hand, what, in like manner, the navy has done? Or what is to be expected from a department not educated in the lights and principles of the new school? The Naval Engineer Corps, which ought to be the soul of the navy, is yet a mere tool to the old school, and destitute of proper organization, and with but a nominal discipline, for true discipline is out of the question where the superior officers are disciples of an obsolete school.

Our new naval warfare is an engineering operation which requires special education and a well-organized corps of engineers, with the distinguishing rank due to that office. The efficiency of the navy is at the mercy of the engineers, and cannot possibly be maintained without due respect to that body.

As it now stands the naval engineer, although in a restricted insignificant position, can manage and manipulate the old school to suit his own personal interest and convenience. He has none above him to fear. His superior officers must trust his words oracularly, of which we have plenty of examples in the navy. An engineer can argue that he has based his operation on physical laws discovered by himself, without being requested to explain such laws. Could that be so if education had done her proper work? or can we find such a case in the Departments of Ordnance, Fortification, or in the Coast Survey?

The country abounds in engineering talent, of which there is plenty in the navy, but it is very rarely much developed, and still less noticed. It would be surprising were it otherwise. The old naval school is not qualified to select or appreciate the engineer of ability; but, even if noticed, he is necessarily doomed to an insignificant rank and a discouraging career. We have the result now before us—the rebel pirate Alabama and others sweeping our vessels from the seas; the numerous blockade-runners can make regular trips in the midst of our boasting navy, on which we have spent hundreds of millions of dollars.

We admit that the naval operations are more difficult, and require manifold more science and talent than those of the army, but, in consequence, they require a corresponding culture in the officers in charge to enable them to bring such physical laws into action, as are involved in the success. This can be attained only by adequate institutions for their education in the magnificent combination of our new naval school, which can only be properly estimated by being perfectly understood. Having now taken the lead of the world in naval warfare, and being unable to derive from foreign sources, either by precept or example, the means of giving a proper organization to the navy, we must follow up and avail ourselves of our own hard-earned experience for this purpose.

In our yet very feeble conception of the importance and of the *range* of knowledge in steam-engineering, it may be suggested that an apartment in the old Naval Academy, to be allotted to steam, would, perhaps, be sufficient for a school of engineers. But then let us reflect for a moment on the immense spectacle now before us, of our growing fleet of ironclads and colossal navy yards, entailing such an unbounded expense, and all of the great interests thus involved and confided with our national reputation to the hands of the engineers, and we must perceive that the latter must not only be equipped with proper knowledge, but must command all the respect and confidence which naturally attach to their important office. It is therefore necessary that a technological school, designed for the special purpose, should be instituted, of the highest possible order, and not limited only to scientific attainment, but foremost in the general *application of the sciences.*

We have numerous examples in Europe, particularly in Russia, where engineers are educated to only scientific attainments, and who, when they enter a machine-shop or engine room, are incompetent for the proper conception of work, but are, nevertheless, intrusted with responsible stations where their practical achievements only lead to mischief.

Our experience throughout life teaches us that a practical man without science seldom makes such serious blunders as a scientific man without practice. The merit then of the *Technonaval Academy* would be in the education of engineers in the practice, and not with mere scientific precepts of professors.

The writer has often observed the career of students from colleges, and regrets to say that

4

too few of them turn their attention to work. Those who have received scientific education generally prefer to become professors, scientific advocates, patent agents, lawyers, philosophical secretaries, &c. &c., whilst the practical operation of our workshops suffer in the extreme. Every once in a while we have a steam-boiler explosion, killing off a great number of men, with great destruction of property; we build vessels which will not float; are often disappointed in the performance of vessels and machinery; we waste great amounts of fuel, and we make extensive and costly experiments in steam-engineering without consulting the physical laws involved in the operation.

In iron foundries castings are often made with too little metal, and sometimes too much; the hydrostatic action of the fluid cast-iron in the mould is rarely understood; the law of shrinkage, strain, direction of crystallization, and sinking, in castings of irregular form, is not generally comprehended; and many other defects of experience exist which often cause the loss of valuable castings, for want of applied science. When the casting turns out a failure, it is generally said that the foundry superintendent is not skilful, or has not experience enough, which often means that he has not made blunders enough to secure success.

The general impression about the business of moulding and casting, as well as all other branches of mechanic arts, *is*, as has been repeatedly told to the writer, namely, that "the profession cannot be brought within the scope of science, but must be learned by experience alone."

On the other hand, scientific men without technical education, intrusted with practical problems, are generally not familiar with important circumstances involved in the operation, which accordingly results in blunders; they are then derided as "scientific men."

Locomotive engineers still allow their thunderbolt to blow out smoke and fire to suffocate passengers, and set fire to houses and forests, when this nuisance of smoke and sparks could be so beneficially utilized in the work for which that fuel is intended.

The combustion of fuel and the utilization of heat, in our present locomotives, are a disgraceful and barbarous abuse of physical laws. The firebox in a locomotive dissolves many times the amount of fuel realized in work, and the heat there generated is so great that it is difficult to find materials for the firebox that are able to withstand it; whilst on the other end of the locomotive, there is applied an arrangement to create a vacuum by the exhaust steam from the cylinders, which, in fact, is only a cooling operation which puts the fire out when it enters the tubes, and the unconsumed carbon, in the form of smoke, with sparks of fire, is blown out through the chimney. The area of the fire-tubes in a locomotive is many times greater than that of the firebox, whilst the evaporative power of the firebox is many times greater than that of the tubes.

We are in possession of sciences, requisite for a more proper arrangement in a locomotive and in steam-boilers generally, by which not only all the carbon could be consumed, but also to utilize the heat in work, but we have not sufficient technical knowledge for their judicious application.

The writer has now gone very far in criticizing our standing as engineers, but hopes to be understood that his motive in so doing is a solicitude for the general interest. There are too many among us to boast of and exaggerate our numerous and real advantages over other nations, but apparently none to point out to us our deficiences.

JOHN W. NYSTROM,

Engineer.

COMMUNICATION

TO THE

SECRETARY OF THE NAVY.

Navy Department,
Washington, D. C., June 26, 1865.

Sir: On the 14th of this month I made an application to the Navy Department for orders to prepare, for the Naval Engineer Cadets, a "course of shipbuilding" based on the "parabolic principle," as explained in the printed papers accompanying the application.

On the 15th I received your communication stating that the proposition was declined.

This hasty refusal of the Department has encouraged me—with all deference—to renew the application in a more specific form, under a conviction that I had probably failed to submit the subject in its true light; inasmuch as upon any other hypothesis, its intrinsic importance could scarcely have failed to have insured for it a different disposition. Indeed, I would deem

4*

it a neglect of duty on my part to drop a subject of such great moment, merely in consequence of a̅ hasty and apparently inconsiderate
reply of the Department, particularly as so many
circumstances may have conspired, amidst a
press of business, to prevent its receiving your
Excellency's personal attention.

When it is considered not only that shipbuilding has not yet been developed to the
condition of a true science, but that the country
contains no school where even the empirical
system under which the construction of vessels
is now carried on, can be learned, we cannot
expect our present naval constructors to dip
into and approve a science which they do not
understand, but distrust and perhaps fear.

By means of long practice and experience,
builders generally attain great skill and taste
in the construction of ships, but, at the same
time, they are yet ignorant of the physical laws
and scientific principles which govern their
success. Their skill and valuable knowledge
die with them, and their toiling successors
must reiterate the same blunders, and gain experience by the same renewed experiments and
errors.

Such will not be the case when shipbuilding
becomes really a science, for then we will be

able to demonstrate, record, and perpetuate
through the unerring aid of mathematical
formulas, a knowledge of the physical laws
which relate to the subject, the accomplishment
of which is now impossible, and the art itself
kept by shipbuilders in profound mystery.

The indifference or hostility of shipbuilders
to a scientific treatment of the subject, arises
mainly from their conviction that it is impossible to bring their profession within the scope
of science, and they persevere in regarding it
as a mere craft. In verification of this, I can
refer to numerous examples, of which the following is one:—

Last summer I made efforts to bring the
" parabolic construction of ships" under the
appreciation of Mr. John Lenthal, the Chief of
the Bureau of Construction, which failed not
only to secure his approval, but absolutely met
with his disparagement. Mr. Lenthal was ap-
prised that the " parabolic system" embodied a
very simple method of recording the peculiarities of vessels, which would be of great importance; he refused, however, to credit the possi-
bility of my plan, and gave me to understand
(without looking at it) that he had all the records which could be necessary already, and
that nothing more was wanted.

Soon after, I was set to work by Mr. Isher-wood, the Chief of the Bureau of Steam-Engineering, to calculate from a great number of ships' drawings these very data, which were not, but which ought to have been calculated and recorded in the Bureau of Construction, and which data are of great importance in questions of steamship performance. In fact the engineer cannot do well without them.

This proves conclusively that the naval constructor was in error regarding the perfection of his records.

There is not to my knowledge an engineer in the naval service who is competent to undertake such an investigation of the properties of ships as that made by me, although the subject belongs directly to his profession. The Corps of Naval Engineers should take the lead in all those progressive changes which naturally attach to their important office, and particularly in those changes which must eventually take place, namely, to combine the construction of machinery and vessels under one head.

Since the introduction into vessels of steam, and other mechanical contrivances which are daily increasing, the two branches have become so intimately connected, that it would be difficult to trace a line of separation between them. The engineer cannot construct the machinery without a knowledge of the vessel which is to contain it, and the shipbuilder cannot properly construct the vessel without consulting the engineer respecting the machiney, and we may expect what happened when the war broke out, namely, to build vessels wholly of iron, for which our present naval constructors are incompetent.

On account, therefore, of the profession of shipbuilding yet being in an empirical condition, and considered separate and apart from that of steam-engineering, there exists much jealousy between the two interests, which results in discord of action; neither of them taking that careful supervision over the whole arrangement which one controlling mind would do.

In many instances where government's vessels have been built in private establishments, the quality of workmanship has suffered considerably for this very reason, of the superintending naval engineer not being familiar with the construction of ships.

The failure of the light-draft monitors affords a still stronger proof of the necessity of instructing engineers in shipbuilding.

As matters now stand, ship constructors are generally so jealous of their profession, that the engineer can **with** difficulty obtain from them the necessary information to govern **him** in his own department; a jealousy which only indicates ignorance. For if their profession **was** brought to the rank of a true science, it could not be kept in a state of mystery, or as a matter of individual knowledge.

Now, believing that I **have** succeeded in developing shipbuilding to the condition of a true science, I desire to throw it wide open for the benefit of all, like the books of Euclid.

My system embodies the results of many years of labor, now in the form of raw materials, to be converted into tables and drawings for general use.

The details of the undertaking are too great **for a** single individual; operating alone, it would cost me several years to complete them, whilst with assistance from the **Navy** Department, it might be accomplished in a few months.

The great labor consists in calculating the tables, which will extend to some five thousand lines, the combination of which would comprehend the construction of an endless number of vessels.

The nature of the tables are very much like

logarithms; they will equally suit any system of weights and measures, or any language and country, and will give the characteristic peculiarities of vessels at the first glance, such as the displacement; areas of water-lines and cross-sections; location of metacentre and centre of gravity, &c. &c.; and they will also give the most important item, so much sought for by scientific men and shipbuilders, namely, the *mean angle of resistance of vessels.*

The displacement of a vessel, bounded within a given *length, breadth*, and *depth*, can vary twenty-five per cent., for the same resistance, in moving through water; a circumstance showing the immense importance of giving the vessel a proper shape.

Shipbuilders generally, through long practice, approach very near the proper shape or form of lines of vessels, but they also often transgress the sought-for limit, which cannot possibly be determined by mere conjecture. But, by the "parabolic method," the most advantageous forms of lines are ascertained and calculated with the aid of tables, which can be used by constructors without a knowledge of mathematics.

For the accomplishment of the object here proposed, I would respectfully request your

Excellency to select two or more young engineers, and place them at my disposition for the purpose of assisting in the calculation of the tables, and of acquiring a thorough knowledge of the " parabolic construction of ships." This would be a very simple and easy course of introducing the science of shipbuilding into the Corps of Naval Engineers, and the Department will thus be placed in possession of scientific resources, which will forever place it out of reach of many errors, and of rash or ignorant experiments which have heretofore wasted so much of its means and its hopes alike unprofitably.

The mode of constructing ships, at the present day, is most generally accomplished by carving out a model from a piece of wood, by eyesight and conjecture—an operation which is often repeated several times before it happens to attain the desiderated end. In more advanced stages of the art, as in the navy, and in some few private establishments, drawings are made, from which models are also executed; but even then the lines are laid down repeatedly from conjecture, until sufficient approximation to the truth is believed to be attained. In both cases, the operation may be likened to the movements of a blind man walking by himself, whilst by the "parabolic method," the construction is started right at the outset, and thus an intelligent perception of principles and results reaches its conclusion with mathematical accuracy.

In the interest of science, as well as that of the Department over which your Excellency has so successfully presided, I earnestly request that my proposition may be considered with the attention which a subject of such importance deserves, and, in conclusion, would suggest, that whilst the government could lose nothing by granting my request, it would gain an advantage which, once possessed, it would never afterwards relinquish.

I have the honor to remain,
Your Excellency's ob't serv't,
JOHN W. NYSTROM,
Act. Chief Engineer, U. S. Navy.

Hon. GIDEON WELLES,
Secretary of the Navy.

5

MEMORANDUM.

NAVY DEPARTMENT,
WASHINGTON, D. C., July 8, 1865.

TO-DAY I called on Secretary Welles, about my application for assistance to calculate the tables for the "parabolic construction of ships," as expressed in the foregoing letter, when Mr. Welles said that he "could do nothing with it, as Mr. Isherwood does not approve your scheme, but says that there is no novelty in it." I then requested the Secretary to respond to my letter to that effect, which he ordered Assistant Secretary Fox to do, but no answer was received.

Assistant Secretary Fox told me that "this democratic government does not take the lead in matters of this kind, as monarchical governments do, but leaves them for the merchant or civil service."

I offered to show Captain Fox some samples of tables for the "parabolic shipbuilding," but he said he "would have no time to attend to it."

It is true the Navy Department has *not taken the lead* in any matter of progress, but left that for the civil service, but it has taken the lead and made extensive and costly experiments in the anti-expansion question of steam; in the building of vessels which will not float; in experimental researches in steam engineering, extensively expatiated upon in large volumes of books upon which I shall make no comments further than to state that the Navy Department has taken the lead in pointing out to the civil service where *not to follow.*

When the war broke out, the naval constructors were not competent to fulfil the requirements in the new era of naval architecture, and there was no naval engineer with requisite technical education to meet the emergency. The Chief of the Bureau of Construction, I understand, declined having anything to do with iron or armored vessels, and the projects for ironclads were necessarily intrusted to officers of the line, the result of which is well known, and not necessary to mention here, for the object of this writing is not to find fault, or to censure those who may have been inadvertently at fault, but to point out the necessity of taking proper steps to prevent similar occurrences in the future, and to prepare to follow up the progressive times.

In regard to the Chief Engineer Isherwood's saying that "there is no novelty in the parabolic system of shipbuilding," I am justified in taking prompt issue with him, for his scientific education does not extend so far as to enable him to judge whether it embraces novelty or not.

I have, however, good reason to believe, from specific indications in discussion with him, that Mr. Isherwood, in his own mind, really thinks that there *is* novelty in the "parabolic method." What can then constitute the object of the Chief in thwarting the interests and progress of the Corps of Naval Engineers?

PARABOLIC CONSTRUCTION OF SHIPS,

NAVY DEPARTMENT.

THE Parabolic System of constructing ships was originated by the celebrated Swedish naval architect, Chapman, about a century ago, at which period it was well received among ship-builders, but on account of its then incomplete form (restricting conductors to particular shapes), it was gradually abrogated, until no trace could be found of it, even in works on shipbuilding. Mr. Chapman hit upon the fortunate idea that the cross-sections of the displacement of a vessel ought to follow a certain progression, in order to present the least possible resistance when moving through the water. He collected a great many drawings of ships of known good and bad performances, and made the following investigation. On each drawing he transformed the cross-sections of the displacement into rectangles of the same breadth as the greatest

beam of the load-water-line of the vessel;
placed their upper edges in the plan of the
load-water-line, by which he found that the
under edges of the rectangles formed a bottom,
the curve of which were parabolas in ships of
known good performances.

Let the accompanying figure, 1, represent a
ship with the load-water-line, w, dead flat cross-
section $a \, \mathbf{x} \, b$, formed into the rectangle $a \, b \, c \, d$,
and $i \, \theta \, i$ another cross-section formed into a
rectangle $e f g h$, so that the breadth $e f$ is equal
to $a \, b$; then the line $k \, l \, m$, Fig. 1, forming the
bottom of the rectangles, should be a parabola
with the vertex at k, and $k \, o$ the axis of the
abscissa.

Mr. Chapman found that the parabola so obtained did not terminate at the stem n, but fell
a little short at m.   The deviation $m \, n$ was very
small in vessels of his days, but in modern vessels it is more considerable, showing that there
must be a point of inflection p in the curve.
However erroneously we may set out in quest
of an object, experience generally leads us towards correct scientific principles.   In the case
before us, experience has increased the deviation
$m \, n$, and we know that inasmuch as nature ad-
mits of no physical by-laws, the curve cannot
be a plain parabola.   It is this increasing de-

Fig. 1.

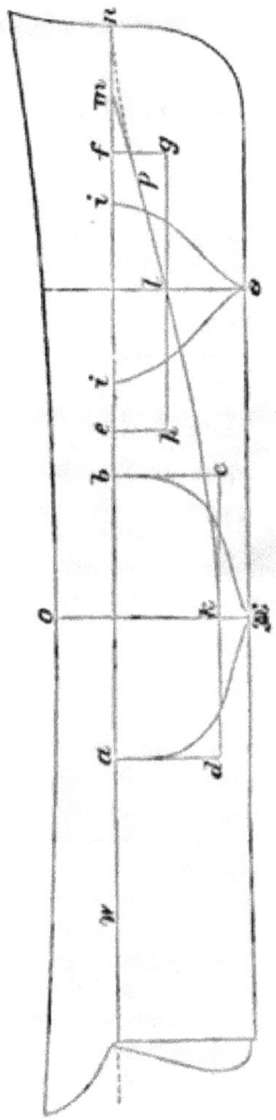

viation *m n* which has led me to investigate the
subject more carefully; starting on the princi-

ple that the resistance to a body in motion in a fluid, is a function of the square of the sine of the angle of incidence to the motion. Let $a\,b\,c\,d$,

Fig. 2.

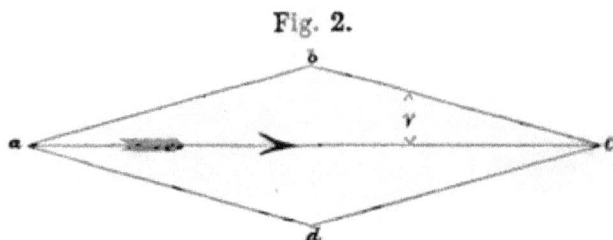

Fig. 2, be a body in motion in a fluid, in the direction $a\,c$; then the resistance to that body is found by experiments to be nearly as the square of the sine of the angle v, omitting friction.

From this it appears that the proper progression of the cross-sections should be as the square of the ordinates in a parabola.

Let Fig. 3 represent a vessel with the dead-flat x and stem n. Draw the cross section $a\,x\,b$, and the rectangle $a\,b\,c\,d$, as before described; draw a parabola $k\,l\,n$ of any desired order, terminating at the stem; then the proper progression of the cross-sections should be as the square of the ordinates β. Let the depth $a\,d =$ 1, then the ordinates β will be fractions of $a\,d$, and the square β^2 multiplied by the area of the dead-flat cross-section x, would give the proper

Fig. 3.

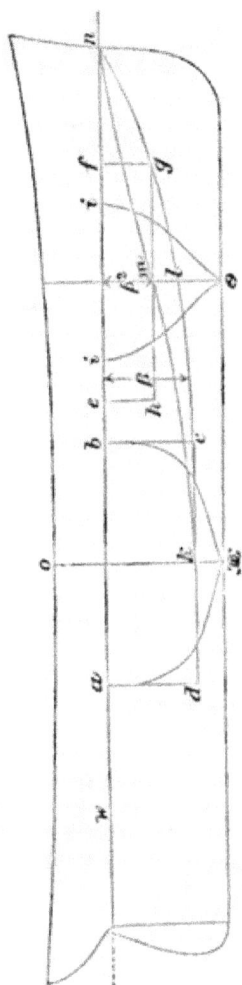

area of the ordinate cross-section θ, or $\theta = \varpi \, \beta^2$,
Fig. 3. The line $k \, m \, n$ should then indicate the

proper progression of the ordinate to cross-
sections θ.   The areas $e\,f\,g\,h = i\,\theta\,i$.

The formula for a parabola in the conic section is—

$$y = \sqrt{2\,p\,x.}$$

Referring to the accompanying figure, o is
the vertex of the parabola, p = parameter, x =
abscissa, and y = ordinate. Applying this for-

Fig. 4.

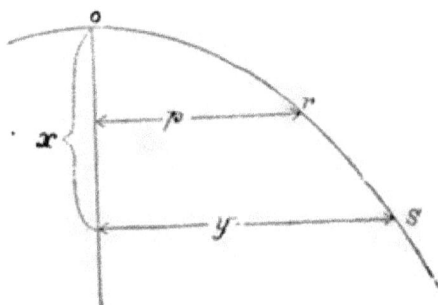

mula and figure to the form of a ship, we place
the vertex of the parabola at the dead-flat x,
the axis of abscissa in the breadth b, and the
largest ordinate y in the length, when the
parabola $o\,r\,s$, Fig. 4, may represent a water-
line in a vessel, as represented in Fig. 5.

The circle, ellipse, parabola, and hyperbola,
in the conic sections are lines of the second order; but in the construction of ships we employ

Fig. 5.

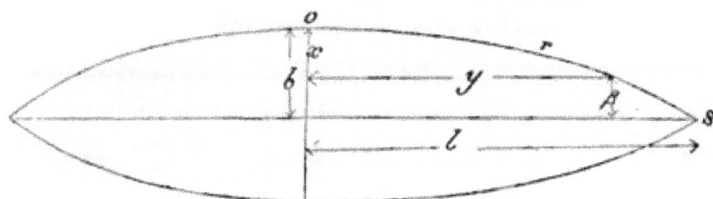

these lines of any order whatever, for which we will denote the index of the root in the parabolic formula by the letter n.

The parameter p is the gauge for the parabola, but is inconvenient for our purpose; it will be better, therefore, to make a gauge that will consist of the given quantities, by limiting the parabolas within the size of the vessel when the limit $x = b$, half the breadth, and the limit $y = l$, half the length of the vessel.

The parabolic formulas will then appear—

$$y = \sqrt[n]{2\,p\,x}, \text{ or } l = \sqrt[n]{2\,p\,b}.$$

Of which—

$$p = \frac{y^n}{2\,x} = \frac{l^n}{2\,b}.$$

$$x = \frac{y^n}{2\,p} = \frac{y^n\,b}{l^n}$$

and—

$$y = \sqrt[n]{\frac{l^n\,x}{b}}.$$

In these formulas the parabola is gauged by the half-length l, and half-breadth b.

Let β denote the distance from the centre-line of a **vessel** to the water-line, then $\beta = b - x$, or $x = b - \beta$, which, inserted in the **above** formulas, will give—

$$x = b - \beta = \frac{y^n b}{l^n},$$

of which—

$$\beta = b - \frac{y^n b}{l^n} = b\left(1 - \frac{y^n}{l^n}\right).$$

Let the depth $a\,d$, figs. 1 and 3, $= b$, represent the area of the dead-flat cross-section x, then the ordinate cross-sections will be—

$$\beta = b\left(1 - \frac{y^n}{l^n}\right) \qquad \cdot \qquad \cdot \quad 1$$

$$\beta = b\left(-1 - \frac{y^n}{l^n}\right)^2 \qquad \cdot \quad \cdot \quad 2$$

The formula 1 gives the plain parabolas $k\,l\,m$, fig. 1; or $k\,l\,n$, fig. 3; or $o\,\beta\,s$, fig. 6; whilst the formula 2 gives the paracyma $k\,m\,n$, fig. 3, or $o\,\theta\,s$, fig. 6.

Formula 1 gives nearly the form of ships **as** constructed in the days of Chapman, whilst formula 2 gives the form of modern ships, constructed for speed.

Fig. 6.

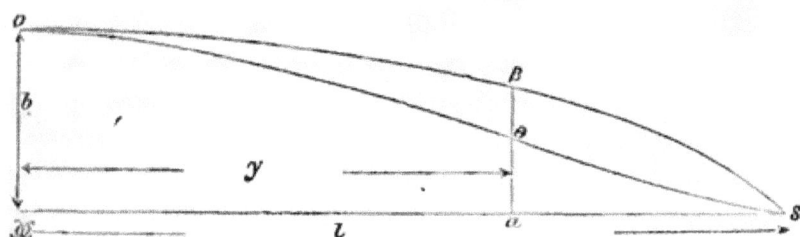

I have investigated the progression of the cross-sections in a great many vessels, from most parts of the world, as will hereafter be shown in a treatise on the parabolic construction of ships now in progress. Many American vessels agree perfectly with formula 2, of which the U. S. frigate *Niagara*, constructed by the late Mr. Steers, is one. The formula 1, which embodies Chapman's method, is therefore not applicable in modern shipbuilding, which I think is the reason why the original parabolic system has not been more generally adopted. It is not always necessary to pay the greatest attention to speed, as there are many other conditions of greater importance, namely, freight, shallow draught, location of metacentre, and centre of gravity of the vessel, for which it becomes necessary so to arrange the parabolic construction of ships, that it will accommodate itself to all the requirements, as well as to the taste of the shipbuilder. This can be accomplished by raising the ordinate β to any arbitrary power, which we will designate with the letter q, and call it the power of the exponent n, when the final formula will appear—

$$\beta = b \left(1 - \frac{y^n}{l^n} \right)^q \qquad . \qquad . \quad 3$$

This is the general formula for the parabolic construction. Simple as it is, it gives any line or form of a ship that can reasonably be required. It will form a square, rectangle, triangle, circle, ellipse, parabola, hyperbola, cyma; all of any order or combination.

Fig. 7.

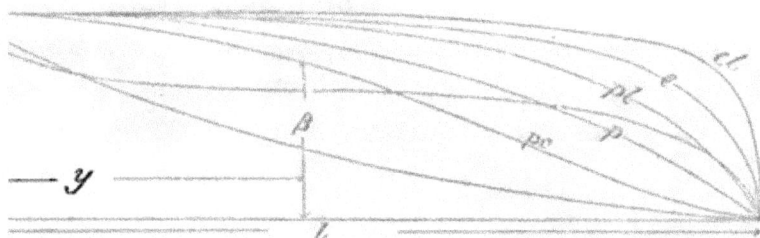

b = half the breadth, or area of dead-flat x.

l = length from x to the stem or stern, or depth under water-line.

For the frames, the depth d, from load water-line to the keel, takes the place of l.

β = ordinate for the line, or ordinate cross-section.

y = abscissa.

n = exponent.

q = power of the exponent n.

The variety of lines represented by fig. 7 are obtained by altering the power q, while n remains constant; or any variety of lines can be obtained for each value of the exponent n.

It is here found necessary to the development of the subject, to propose or establish new names to such lines as have not heretofore been defined or subjected to an algebraic formula. The degree of development of an art may be correctly measured by the perfection of its vocabulary. As the construction of ships has not heretofore been brought to a perfect system, we have not been able to define the great variety of lines or forms of ships. We can say a vessel is very sharp, or very full, with more or less rise of floor; but have no language by which to convey correctly, how sharp, how full, or with how much rise of floor. As an illustration it may be mentioned, that on one occasion I met some shipbuilders, and discussed with them the construction of ships, when one said, "I am constructing a ship that will be so sharp, that you cannot roll a barrel on the lower deck, within fifteen feet of the bow," which made me but little wiser. Now, in the language of the parabolic construction, to convey the same idea with precision and accuracy, we have only to give the exponent and power, which not only impress the mind clearly with the correct degree of sharpness, but also with the complete form of the vessel.

On another occasion I remarked to a shipbuilder, in his yard, that "the lines of a certain vessel were too sharp, and if made fuller it would go much easier through water with considerably more displacement." The shipbuilder acknowledged that it appeared to him to be so, but remarked, with the usual practical sneer, that "it is very easy to see that after the vessel is finished." The great merit in the parabolic construction is, that we know the mechanical and physical properties of the lines before they are laid down; we need not even look at them for such purpose.

We have both in Europe and America many curiously constructed vessels, and some of them reported to perform wonderfully, but we have not been able to record their peculiarities; for even the drawing of their lines would fail to convey with correctness what constitutes their novelty or folly.

It is therefore proposed to establish the following technical terms in naval architecture:—

Any line $p\ l$, in the accompanying figure 7, located between the parabola p and ellipse e, is to be called *Paralipse*.

Any line $p\ c$, located under or within the parabola p, to be called *Paracyma*. In architecture, cymas are generally constructed of circle-arcs, but in this case cymas are derived from parabolas.

Any line $e\ l$, extending outside of the ellipse, to be called *Evolipse*.

In modern constructed vessels, those lines are generally distributed as follows:—

All water-lines of the displacement are *Paracymas*, with the highest power near the keel, approaching parabolas near the load water-line, which latter may also be a *Paracyma*. The frames are generally *Paralipses* about the middle of the vessel, and terminate in *Parabolas* and *Paracymas*, in the stern and bow. Above the water, the horizontal lines are generally *Parabolas* in the foreship; and in the aftership, *Paralipses*, *Ellipses*, and *Evolipses*.

The power q defines the line as follow:—

Parabola p, $\beta = b\left(1 - \dfrac{y^n}{l^n}\right)^q = 1.$

Ellipse e, $\beta = b\left(1 - \dfrac{y^n}{l^n}\right)^q = \dfrac{1}{n}.$

Circle, $\beta = R\left(1 - \dfrac{y^n}{R^n}\right)^q = \dfrac{1}{n}.$

Paracyma $p\,c$, $\beta = b\left(1 - \dfrac{y^n}{l^n}\right)^q > 1.$

Paralipse $p\,l$, $\beta = b\left(1 - \dfrac{y^n}{l^n}\right)^q$ between 1 and $\dfrac{1}{n}.$

Evolipse $e\,l$, $\beta = b\left(1 - \dfrac{y^n}{l^n}\right)^q < \dfrac{1}{n}.$

In my treatise on the parabolic construction of ships, now in progress, there will be calculated 54 values of the power q, each with 90 different exponents n, making 4860 different lines, which will cover the most general requirements in practice. Samples are here given.

SAMPLE TABLE FROM FORMULA 1.

Exp. n.	1	2	3	4	5	6	7	a'N'D'	e'	m'	t'
1	.1250	.2500	.3750	.5090	.625	7500	.8750	.5000	.3333		
1.25	.1237	.3020	.4443	.5795	.7085	.8232	.9297	.5555	.3461		
1.5	.1815	.3505	.5059	.6464	.7701	.8750	.9558	6000	.3571		
1.75	.2084	.3955	.5607	.7027	.8203	.9116	.9737	.6363	.3666		
2	.2344	.4375	.6094	.7500	.8394	.9375	.9844	.6666	.3750	.3021	1.333
2.25	.2595	.4765	.6527	.7898	.8860	.9558	.9907	.6923	.3823	.3285	1.447
2.5	.2838	.5129	.6912	.8232	.9148	.9687	.9945	.7142	.3888	.3502	1.562
2.75	.3073	.5466	.7234	.8513	.9326	.9779	.9967	7333	.3948	.3688	1.682
3	.3390	.5781	.7558	.8750	.9476	.9844	.9988	.7500	.4000	.3857	1.800
3.25	.3521	.6074	.7829	.8949	.9587	.9889	.9988	.7647	.4047	.4010	1.922
3.5	.3733	.6346	.8070	.9116	.9677	.9922	.9993	.7777	.4091	.4144	2.040
3.75	.3939	.6600	.8284	.9256	.9747	.9945	.9996	.7894	.4130	.4265	2.160
4	.4138	.6836	.8474	9375	.9802	.9961	.9998	.8000	.4166	4376	2.285
4.5	.4517	.7260	.8794	.9559	.8979	.9980	.9999	8181	.4230	.4570	2.534
5	.4871	.7627	.9016	.9687	.9926	.9990	.9999	.8333	.4286	.4734	2.775
6	.5512	.8220	.9404	.9841	.9972	.9997	1.000	.8571	.4375	.5000	3.270
7	6073	.8665	.9627	.9922	.9989	.9999	1.000	.8750	.4444	.5202	3.770
8	.6564	8999	.9767	.9961	.9996	.9999	1.000	.8888	.4500	.5354	4.272
12	7986	9683	.9964	.9997	.9999	1.000	1.000	.9231	.4643	.5768	6.275
16	.8819	.9900	.9994	.9998	.9999	1.000	1.000	.9412	.4720	.6060	8.273

SAMPLE TABLE FROM FORMULA 2.

Exp. n.	1	2	3	4	5	6	7	a'N'D'	e'	m'	t'
1	.1563	.625?	1406	.2500	.3906	.5625	.7656	.3333	2500		
1.25	.2363	.912?	1974	.3359	.4992	.6777	.8569	.3968	.2691		
1.5	.3295	.112?	2547	.4179	.5934	.7656	.9136	.4500	.2860		
1.75	.4342	.156?	3144	.4938	.6729	.8310	.9481	.4949	.3000		
2	.5493	.1011	3713	.5625	.7385	.8789	.9690	.5833	.3125	.2273	1.219
2.25	.6735	.2271	4260	.6237	.7920	.9136	.9815	.5664	.3189	.2541	1.273
2.5	.8036	.2050	4777	.6777	.8371	.9385	.9880	.5952	.3333	.2770	1.335
2.75	.9445	.2988	5262	.7248	.8695	.9563	.9934	.6205		.2979	1.402
3	.1090	.3342	5713	.7656	.8980	.9690	.9961	.6428		.3164	1.472
3.25	.1240	.3689	6130	.8008	.9192	.9779	.9977	.6627	.3371	.3255	1.545
3.5	.1394	.4028	6512	.8310	.9365	.9844	.9986	.6805	.3636	.3500	1.630
3.75	.1555	.4356	6862	.8569	.9501	.9890	.9992	.6966	.3895	3668	1.697
4	.1712	.4673	.7184	.8789	.9608	.9922	.9995	.7111	.3750	.3765	1.773
4.5	2045	.5276	.7738	.9137	.9750	.9961	.9998	.7383	.3846	.4000	1.929
5	.2373	.5804	.818?	.9385	.9857	.9980	.9990	.7576	.3929	.4196	2.088
6	.3038	.6729	.884	.9890	.9945	.9995	.9999	.7912	.4062	.4517	2.407
7	.3688	.7568	.926	.9844	.9971	.999?	1.000	.8166	.4166	.4768	2.732
8	.4308	.8099	.954?	.9922	.9992	.999?	1.000	.8366	.4250	.4962	3.060
12	.6377	.937	.992	.9995	.9998	1.000	1.000	.8861	.4464	.5463	4.390
16	.7778	.989	.995	.9999	1.00	1.00?	1.000	.9129	.4629	.5783	5.715

The length l from the dead-flat x to stem or stern, also the draft of water from the load-line to the base-line, are each divided into eight parts, forming the ordinates, 1, 2, 3, 4, 5, 6, 7, in the table, counted from the stem or stern of the centre x, or from base-line to water-line, as represented in the accompanying plate.

Either table, exponent, or power, can be employed for either frames, water-lines, or displacement.

Area of any water-line **a** or a.

" " cross-section x or θ.

Cubic contents of displacement D.

$$\left.\begin{array}{l} \\ \\ \end{array}\right\} = \int \beta\,dy,$$

which integral coefficient is contained in the column $a'x'D'$.

The depth of the centre of gravity of any cross-section, or of the displacement, or the distance from the dead-flat x, to the centre of gravity of the area of any water-line, or of the fore or aft part of the displacement, will be—

$$- \quad e = \int \frac{\beta\, y\, dy}{a,\ x,\ \text{or D}},$$

which integral coefficient is contained in the column e'.

The height of the metacentre will be—

$$m = \frac{2}{3D} \int \beta^3\, dy,$$

which integral coefficient—

$$m' = \frac{2}{3}\int s^3\, ay,$$

is contained in the column m'.

When the power q and exponent n are given, we have—

Height of metacentre $m = \dfrac{l\, b^3\, m'}{D}$.

Momentum of stability $=$ Q sin. $v\left(\dfrac{l\, b^3\, m'}{D} \pm g\right).$

Q $=$ weight of the vessel, and $g =$ vertical height between the two centres of gravity.

The mean angle of resistance of the vessel through water is found by the following formula:—

$$tang.\ v = q^2\, b\, n^2 \int \left(1 - \frac{y^n}{l^n}\right)^{2q-2} \frac{y^{2n-2}}{l^{2n}} dy.$$

The integral coefficient of this formula is contained in the last column t' in the table.

It does not appear that Chapman attempted to form the water-lines and frames of a vessel by the parabolic method. He says the area of the cross-sections can be approximated by a parabola, placing the vertex at the keel; but this cannot give a proper shape to the frame. Inasmuch as the displacement of a vessel is the integral of the areas of the water-lines and

cross-sections, and as those areas are integrals
of the ordinates in the frames and water-lines,
they are all convertible into one another by a
common formula, which is the formula 3, and
which formula, simply by placing $q = 1$, embodies Chapman's system completely. But by
so doing, the constructor is restricted to a
stiff and obstinate guide, which will not yield
to his taste, and we have the result before us;
namely, the shipbuilder assumes his independence. It would be futile to attempt to introduce a system of constructing ships that would
not accommodate itself to the taste of the constructor. By Chapman's system, when the
length, breadth, depth, and the displacement
are given, then the sharpness of the vessel is
obdurately fixed; while by giving an arbitrary
value (as here proposed) to q, the sharpness
and ease of the lines can be made to vary considerably, and accommodate themselves to the
taste of the architect.

Suppose the area, length, and breadth of the
load-water-line of a vessel are given, which is
substantially the same as if the displacement,
dead-flat, cross-section and length were given;
then Chapman's method, formula 1, will produce the fixed line, say $o\,m\,m\,s$, fig. 8, while the
formula 2 will produce any variety of lines, as

o o o s, or *o e e s,* or if we wish to go to the extreme the wrong way, we can produce the line *o n n s;* in fact, the formula 3 can manipulate the displacement the same as one can work a lump of soft clay in his hands. This is a pro-

Fig. 8.

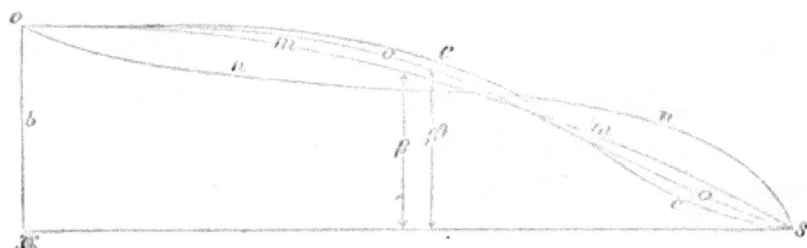

perty of my parabolic system which does not yet appear to have been appreciated, but whose utility, when once fairly understood, must be universally accepted.

It is not to be supposed that this short outline of the parabolic construction embodies the full capacity of that method, for which a much more complete work would be required. When constructors become accustomed to the tables, they can readily select the proper exponents, and reason intelligibly with each other on the forms of lines and vessels.

APPLICATION.

Let it be required to construct a vessel of the following dimensions:—

Length in the load-line,	$L = 325$ feet.
Breadth of **beam**,	$B = 40$ "
Draft of water from base-line,	$d = 18$ "

Let the dead-flat cross-section be selected from table 1, of the exponent $n = 6$; then the numbers in the line 6, multiplied by half the beam $b = 20$ feet, will give the corresponding ordinates in the dead-flat frame; and the area will be $\mathcal{K} = B \, d \, \mathcal{K}' = 40 \times 18 \times 0.8571 = 617.112$ square feet.

Let the load-water-line be selected from table 2, of exponent $n = 3$; then the numbers in the line 3, multiplied by half the beam $b = 20$ feet, gives the corresponding ordinates in the water-line, and the area will be $a = L \, B \, a' = 325 \times 40 \times 0.6428 = 8356.4$ square feet.

Let the displacement be selected from table 2, and of exponent $n = 3.25$, then the numbers in the line 3.35, multiplied by the dead-flat cross-section $\mathcal{K} = 617.112$ square feet, will give the corresponding ordinate cross-sections of the displacement. The cubic contents of the displacement will be $D = \varpi L D' = 617.112 \times 325 \times 0.6627$ $= 132912$ cubic feet, or 3797.4 tons.

The sample tables here given do not extend so far as to allow a correct calculation of the depth of the centre of gravity of the displacement, but suppose the areas of the water-lines to progress with the exponent $n = 5$, table 1, then the depth of the centre of gravity of the displacement will be $de' = 18 \times 4286 = 7.7148$ feet.

The height of metacentre above the centre of gravity of the displacement will be—

$$m = \frac{L\, b^3\, m'}{D} = \frac{325 \times 20^3 \times 0.3164}{132912} = 6.1898 \text{ feet.}$$

This metacentre is very low on account of having assumed a very sharp water-line.

The tangent for the mean angle of resistance will be—

$$tang.\ v = \frac{\varpi\, t'}{L\, d} = \frac{617.112 \times 1.472}{\frac{1}{2} \times 325 \times 18} = 0.31056$$
$$= tang.\ 11° 33'.$$

The actual resistance by impact and friction, the wet area of the hull of the displacement, &c. &c., are calculated by simple formulas not given in this short outline of the parabolic construction.

7

Recording Formula.

The form of any vessel may be recorded by one general formula, as follows:—

$$\left.\begin{array}{l} \text{W}\,nq \\ \text{D}\,nq \end{array}\right\} \; l \left(\begin{array}{l} \text{L}\,b\,d \\ \text{æ}\,nq \end{array}\right) l \left\{\begin{array}{l} nq. \\ nq. \end{array}\right.$$

The first factor $\left.\begin{array}{l} \text{W}\,nq \\ \text{D}\,nq \end{array}\right\} l$ represents the properties of the after-body of the displacement. W nq represents the exponent n and power q of the load-water-line; Dnq the exponent n and power q of the displacement, and l the length from the stern-post to the dead-flat æ in a fraction of the whole length L of the vessel.

The second factor $\left(\begin{array}{l} \text{L}\,b\,d \\ \text{æ}\,nq \end{array}\right)$ represents the dead-flat, or L the whole length of the vessel, b the half dead-flat breadth in the load-water-line, and d the drift of water from the base-line to the load-water-line. ænq represents the exponent n and power q of the dead-flat æ.

The third factor $l \left\{\begin{array}{l} nq. \\ nq. \end{array}\right.$ represents the properties of the fore-body of the displacement; l is the length from the dead-flat æ to the stem of the vessel; n and q represent the exponents n and powers q for the load-water-line and the displacement respectively; l and l may be expressed in real length, as feet.

A well-proportioned sailing yacht may be set up as follows, with numerical values in the general formula:—

$$\begin{matrix} \text{w} \, 3 \times 2 \\ \text{D} \, 2 \times 2 \end{matrix} \Big\} \; 0.381 \left(\frac{80 \times 8 \times 8}{\pi \, 3 \times 4} \right) 0.619 \; \Big\{ \begin{matrix} 2.75 \times 2. \\ 2 \times 2. \end{matrix}$$

These data will enable a shipbuilder to construct a sailing yacht of definite shape.

When constructors become accustomed to the parabolic method, they can determine with great correctness the exponent n and power q of any line at first sight, and thus enable them to record by the above formula, the form of any vessel exposed to view, from which a similar vessel can afterwards be constructed.

Records of this kind have been frequently made in shipyards by the author, of which a case may be mentioned, namely, the "Dictator" (built by Hogan & Delamater, New York), of which the following formula was recorded:—

Formula for the Dictator.

$$\begin{matrix} \text{w} \, 5.5 \times 1.75 \\ \text{D} \, 4.75 \times 3.25 \end{matrix} \Big\} \; 96 \left(\frac{240 \times 21 \times 16}{\pi \, 2.75 \times 0.5} \right) 144 \; \Big\{ \begin{matrix} 2.75 \times 1.5 \\ 3 \times 2. \end{matrix}$$

From these data a vessel can, at any time, be constructed similar to the "Dictator," by any one familiar with the method. The draft 16 feet is from the base-line to the under side of guards, but she draws some four feet more water.

A skilful shipbuilder may, by his empirical mode of reasoning, be able to memorize, for a short time, the form of a ship, but most likely in a clouded condition, which will soon vanish away.

Recording Tables.

The properties of a vessel can be more minutely and fully recorded in a table of a general form, as follows. The vessel being divided from the dead-flat x to the stern and to the stem, also from the base-line to the load-water-line, into eight equal parts, as shown on · the accompanying plate:—construct the following two tables, one for the after-ship and one for the fore-ship. The data given in these tables are for a steamboat. The top line s contains the sheer of the vessel, or height from the load-water-line to the rail, at each division or ordinate. The line R contains the ordinates for the rail, and D that of the deck. In these tables, the dimensions do not correspond with those on the plate. The line D w means a line of a plane tangenting the deck at x, and parallel to the load-water-line; the deck-line was not shown on the drawing from which this table is

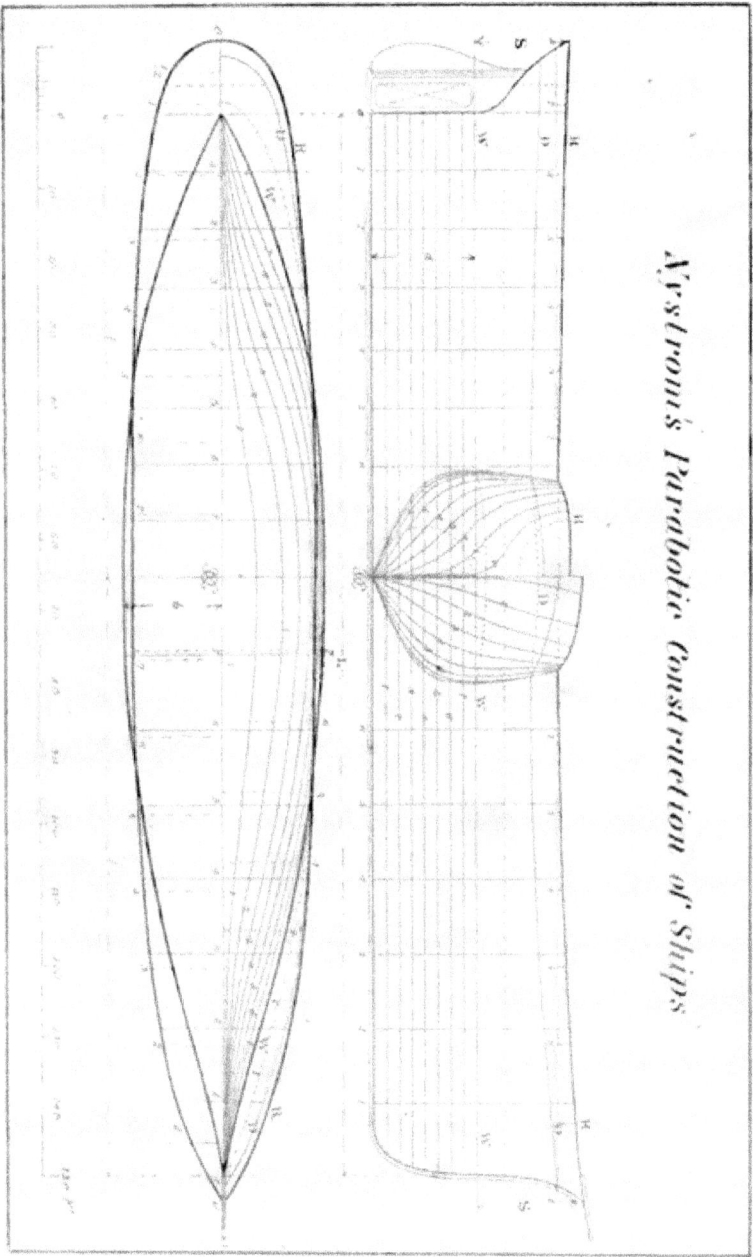

Nystrom's Parabolic Construction of Ships

made. The line *Ord.* contains the number of each ordinate from stem or stern to the dead-flat x. The line w contains the ordinates for the load-water-line, and the lines 7, 6, &c. &c., contain the ordinates for the corresponding water-lines.

The line *o* contains the half-width of throat in the base-line. The line θ contains the half-areas of the ordinate cross-sections of the displacement. The line n contains the exponent of the frames. The line q contains the powers of the frames. The line u contains the length of each frame from the base-line to the load-water-line.

The column **a** contains the area of each water-line, n the exponent, and q the power for the corresponding water-line. Column *o* contains the half-width of throat on the stern-post or stem; in this case it shows that the boat is a propeller, because the throat is widest at the ordinate 4, where the propeller shaft goes through the stern-post. Columns **A** or **F** contain the ordinates for the throat on stern-post or stem, measured from the perpendicular. Column u contains the length of the corresponding water-line from stem or stern to the dead-flat x.

The corner θ x contains the half area of the greatest immersed cross-section, which, in this

7*

case, is 133.4 square feet. Corner θ a contains the half-displacement of the vessel from x to stem 10000, or stern 8528 cubic feet. The corners θ n and θ q contain the exponent and power of the displacement longitudinally; and the corners a n and a q contain the exponent and power of the displacement vertically. The corner q q contains the mean angle of resistance 13° 54', or mean angle of delivery 14° 49'. The corner u u contains the wet surface of half the displacement from x to stern or stem. The constructing draft of water and length are contained in the corners o u and u o.

This form of table will suit for any shape or size of vessel. It is like a tailor's measurement of a coat. When the shipbuilder becomes ac-customed to it, he can see, at the first glance, the properties of the vessel.

When thus brought to a system, forms of tables could be printed and bound in a book, for the use of shipbuilders.

The general formula for this steamboat is—

$$\begin{matrix} \text{w} & 3.375 \times 0.824 \\ \text{D} & 3.375 \times 1.375 \end{matrix} \Big\} \ 88.6 \ \left(\frac{188.6 \times 15.35 \times 10}{x\ 6.75 \times 1.125} \right) 100 \ \Big\} \ \begin{matrix} 4.5 \times 1.25. \\ 3.625 \times 1.375. \end{matrix}$$

When the shape of the vessel is thus ob-tained and recorded, divide the frames as re-quired in building the ship.

The formula for the steam-propellor repre-sented on the plate is—

$$\begin{matrix} \text{w} & 2.5 \times 1 \\ \text{D} & 2 \times 2 \end{matrix} \Big\} \ 65.625 \ \left(\frac{150 \times 15 \times 15}{x\ 3 \times 1} \right) 84.375 \ \Big\} \ \begin{matrix} 2 \times 1.18. \\ 2 \times 2. \end{matrix}$$

Records of a Steamboat.—After-Ship.

Ord.	0	1	2	3	4	5	6	7	w̄	a	z	q	A	u
S	11.60	10.90	10.30	9.83	9.50	9.40	9.32	9.25	9.20				15.00	109.6
R	0.00	11.22	12.34	12.98	13.50	13.95	14.20	14.35	14.40					101.6
Dw	0	6.95	11.20	13.11	14.05	14.72	14.90	15.06	15.15				8.88	
W	0.50	6.30	10.30	12.70	14.10	14.90	15.15	15.30	15.35	1084	3.375	0.824	0.00	91.62
7	0.50	5.20	9.26	12.18	13.90	14.88	15.14	15.28	15.34	1040	3.6	1.12	0.80	91.40
6	0.60	4.33	8.30	11.52	13.63	14.75	15.11	15.28	15.32	1000	3.5	1.3	1.27	91.00
5	0.78	3.63	7.35	10.80	13.21	14.58	15.08	15.25	15.30	976	3.25	1.375	1.35	90.75
4	0.85	3.12	6.55	9.90	12.60	14.23	14.89	15.12	15.22	935	3	1.4	1.41	90.00
3	0.80	2.54	5.62	8.80	11.60	13.50	14.42	14.60	14.91	884	2.75	1.5	1.48	89.55
2	0.60	1.95	4.51	7.40	10.09	12.20	13.20	13.60	13.90	772	2.575	1.75	1.50	89.15
1	0.50	1.20	2.90	4.90	6.45	7.50	8.10	8.35	8.50	.507	2.68	1.625	1.50	88.91
0	0.50	0.50	0.50	0.50	0.50	0.50	0.50	0.50	0.50				8.61	88.60
9	0.00	31.50	63.70	92.00	114.7	123.9	129.3	131.0	133.4	8528	3.375	1.375		811.0
w̄	10.00	5625	.9375	1.187	1.800	4.250	6.25	6.50	6.75	1.687	17.84	14°49'		90.14
q		0.611	0.375	0.422	0.368	0.808	1.125	1.125	1.125	0.397				
u		12.13	15.00	17.20	19.30	20.85	21.70	21.62	22.30	160.57				1607.5 sq. ft.

From w to D w, 4 feet.   From D to R, 3.5 feet.

Records of a Steamboat.—Fore-Ship.

	0	1	2	3	4	5	6	7	$\overline{\omega}$	a	n	q	p	n
S	12.50	11.74	11.15	10.65	10.20	9.87	9.66	9.35	9.20				-1.20	105.4
R	1.40	10.16	12.68	14.44	14.59	14.50	14.50	14.49	14.40				-0.1	103.0
Dw		6.75	11.35	13.78	14.80	15.02	15.07	15.10	15.15					
Ord.	0	1	2	3	4	5	6	7	$\overline{\omega}$	a	n	q	p	n
W	0.50	5.60	10.25	13.10	14.50	15.20	15.27	15.30	15.35				1.25	102.8
7	0.50	5.14	9.90	12.87	14.33	15.16	15.25	15.29	15.34	1218	4.5	1.25	1.75	102.8
6	0.50	4.72	9.46	12.55	14.14	15.08	15.22	15.28	15.32	1200	4.5	1.375	2.25	102.9
5	0.50	4.25	8.83	12.04	13.81	14.92	15.16	15.25	15.30	1184	4.25	1.375	3.00	103.0
4	0.50	3.75	8.09	11.35	13.31	14.62	14.90	15.12	15.22	1158	3.875	1.375	4.00	102.6
3	0.50	2.97	7.12	10.30	12.50	14.11	14.40	14.60	14.91	1102	3.56	1.50	5.15	101.2
2	0.50	2.00	5.73	8.50	10.95	12.70	13.19	13.60	13.90	1053	3.25	1.50	6.80	100.9
1	0.50	0.75	3.30	4.80	7.30	7.50	8.00	8.35	8.50	925.5	2.875	1.625	9.15	100.3
0	0.50	0.50	0.50	0.50	0.50	0.50	0.50	0.50	0.50	572.5	3.125	1.75	17.00	100.0
θ		34.8	72.15	98.2	119.1	125.5	129.3	131.2	133.4	10000	3.625	1.375		9165
n	10.00	1.35	1.875	2.72	2.75	4.75	5.5	6.5	6.75	1.5	18.03		18354 sq. ft.	101.8
q		0.75	0.766	0.845	9.33	1.00	1.00	1.125	1.125	0.222		13°54'		
n		11.00	15.45	18.20	20.15	21.40	21.80	22.00	22.30	162.30				

The general shipbuilding tables from formula 3, have been calculated, corrected, and rearranged several times. The undertaking is similar to that of calculating tables of logarithms, and although equally extensive, is much more complicated, and too great a task for a single individual. The immense amount of labor which has been spent on logarithms by different mathematicians, in different countries, is well known, as also that it required some two centuries before they were brought to a condition of thorough reliability.

Baron Napier invented the foundation of logarithms printed in his *Canon Mirabilis Logarithmorum*, in the year 1614, but started on an inconvenient basis, which was improved by Professor Henry Briggs in 1615, who calculated our present common logarithms for the natural numbers up to 30,000, and in 1628, the logarithms for all natural numbers were computed for the first time up to 100,000. Since then the logarithms have been calculated over and over again by different mathematicians, who have continually discovered errors in the same, until very recently the last edition of Vega's tables has been generally accepted as correct.

The shipbuilding tables may be considered in a similar situation.

The difficulty is first to calculate a complete and well-arranged set of tables, then to have them perfectly corrected and purged from errors; all of which could have been accomplished in the Bureau of Steam-Engineering, had I only succeeded in securing for them the appreciation of the Navy Department.

The individual sacrifice of labor and time necessary to perfect these tables is altogether inconceivable by the uninitiated, and would never be compensated by the immediate sales of such a publication. I, therefore, amidst my multifarious and pressing engagements, leave to others both the profit and distinction that may accrue from their ultimate perfection, and will cheerfully contribute my quota as a purchaser, to their cost, rather than assume this herculean labor myself.

CONSTRUCTOR'S OFFICE, U. S. NAVY YARD,
PHILADELPHIA, Sept. 14, 1865.

SIR : I have examined your proposed method of constructing ships, called the parabolic construction, and am of the opinion it will be very useful for the shipbuilding profession, and think it embraces, in full, the merit therein described.

I am, very respectfully,

Your obedient servant,

W. L. HANSCOM,

Naval Constructor.

J. W. NYSTROM,

Civil Engineer,

Philadelphia.

PHILADELPHIA, Sept. 14, 1865.

SIR: Having been shown the system proposed by you for calculating the data necessary
in the construction of the models of vessels, I
am of the opinion that so certain and easy a
mode of ascertaining the shape and dimensions
hitherto assumed by individual judgment,
would be immensely valuable to the profession.

And with regard to the principle on which
said system is based, I have no reason, from
my present knowledge of it, to doubt that by
its adoption, at least a great improvement in
models over the average now made would
result.

Very respectfully,

J. VAUGHAN MERRICK.

J. W. NYSTROM, ESQ.

RESIGNATION.

NAVY DEPARTMENT,
WASHINGTON, D. C., July 8, 1865.

SIR: I respectfully beg leave to tender my resignation as Acting Chief Engineer in the U. S. Navy.

At the time the Navy Department paid me the compliment of declining a previous resignation (tendered on the 3d of February last), I gave as a reason that "the pay was much less than I could obtain in private employment, whilst the living was much higher in Washington."

However true this may have been, the real reason for tendering my resignation, both then and now, as stated to Mr. Isherwood, the Chief of the Bureau of Steam-Engineering, was that I have failed in bringing my attainments and qualifications to the notice and due appreciation of the Department.

My present duties are limited, I may say, to questions of simple arithmetic, which could be

8

performed by a schoolboy, whilst my engineering knowledge, **which is** actually needed, and could **be** advantageously **employed** for **the** benefit **of** the navy and the country, **is** thrown **away.**

Under **a conviction** that my knowledge of naval engineering would render me eminently **useful,** and that from the present condition **of** the country **the** Department actually **requires the** utilization **of every** possible means which **could be** directed **to the advancement** of this paramount **interest, I cannot** conscientiously continue **on** the **pay-rolls of the** navy whilst **the** class of services I am required to render are at once so unworthy of myself, and so inadequate a requital to the government for the emolument which **it** so generously confers upon me.

I have the honor to remain,

Your Excellency's ob't serv't,

JOHN W. NYSTROM,

Act. Chief Engineer, U. S. N.

Hon. GIDEON WELLES,
Secretary of the Navy.

MEMORANDUM.

WASHINGTON, July 10, 1865.

THE acceptance of my resignation, tendered on the 8th inst., was received this morning.

I have thus resigned a position in the navy where my professional attainments are most needed, and where my engineering knowledge could not be utilized because there was no one in the Navy Department who could appreciate or employ them. There is work in the Bureau of Steam-Engineering for a dozen engineers of my qualifications, and there are now many good mathematicians in that Bureau who would be very glad to undertake such work as is now needed in the organization and instruction of the Engineer Cadets, but there is no one in the Department with adequate technical knowledge to take the lead in such an important enterprise.

The Navy Department is apparently unaware that our present scientific books are not only inadequate to meet the requirements of the

day, but much of the matter existing in them
is very confused, without order or classification,
and some of it is not correct. Besides, many
of our scientific books contain an unnecessary
burden for students.

In the multifarious studies required in our
days by Naval Engineers, it is of great importance to economize their time and labor.

The science of dynamics is yet in a very
complicated and confused condition, without a
specific meaning being attached to the terms
employed. Correspondents are constantly
seeking information, through scientific journals, on the subject of dynamics, and invariably
receive confused answers, in verification of
which a few examples may be given.

The *Scientific American*, of November 26,
1864, informs its correspondents that "the size
"and weight of a fly-wheel must be in proper
"proportion to the machine which it is designed
"to regulate, and this is determined by observa-
"tion and experience; it cannot be calculated
"by any mathematical rule. Within the limit
"usually adopted by mechanics, our preference
"is for light wheels of large diameter, rather
"than for heavier ones of smaller diameter. The
"regulating power of fly-wheels is in proportion

"to their weight multiplied by the square of
"their velocity."

Here it is asserted that the weight of a fly-
wheel must be in proper proportion to the
machine! whilst we know that many machines
run well without a fly-wheel; its sole function
and office is to approach uniformity of motion
by regulating alternate irregular work. The
Scientific American says that "the proper size of
the fly-wheel cannot be calculated." The action
of a fly-wheel, however, is calculated and determined as easily as a simple problem in geometry; but we have yet no books where this
is properly explained.

An English scientific journal informs its correspondent that "the size and weight of fly-
"wheels are usually determined from practical
"experiment. There is given an elaborate
"theory of the fly-wheel in Moseley's *Mechanical
"Principles of Architecture and Engineering*, but
"the formulas deduced are very intricate."

The formulas here quoted are complicated,
because the subject is not properly understood.

The *Scientific American* also for Sept. 10, 1864,
informs its correspondent that "when a body
"is raised slowly, the power required to over-
"come the inertia is inappreciable, and must be

"disregarded in reckoning the work done. But
"when the velocity is appreciable, it must be
"considered in computing the work. This part
"of the work is in proportion to the square of
"the velocity." Now the work required to raise
a body is equal to the weight of the body multiplied by the height to which it is raised, independently of velocity. The work expended on
the inertia in starting the body, is re-utilized
when it is brought to rest. From the *Scientific
American*, we may infer that work is required
to overcome the inertia, while the body is raised
with a uniform velocity.

Another correspondent is informed that "the
"vis-viva, or force of a moving body, is in pro-
"portion to the square of the velocity, and the
"power required to impart velocity is in the
"same ratio. It therefore requires an expendi-
"ture of four times the force to impart double
"velocity either to a projectile or to a revolving
"wheel."

Here *force, power*, and *work* are, as usual,
confounded with each other. A force of one
pound can give as much velocity to a body free
to move as a force of a hundred pounds, if
time be disregarded. In equal times, the force
is directly as the velocity. In equal space, the

force is as the square of the velocity, or inversely as the square of the time.

The science of dynamics is yet in the condition which geometry would be without illustrations. Dynamical quantities are physical operations, and cannot be recognized as material or geometrical objects, but must be conceived from algebraical formulas. But among the very best mathematicians, there are few who can conceive the true configuration of an object when it is simply expressed in a complicated formula. Dynamical quantities, such as *force, velocity*, and *time*, and their combinations, into *power, space*, and *work*, can be compared with and illustrated by geometrical objects, and thus made to present a clear conception to the mind, without which it is often difficult, if not impossible. At least, I have not myself been able to form a clear conception of dynamics without the aid of adequate illustrations.

It yet remains to explain and illustrate how work is accumulated in, and distributed by a fly-wheel; how the combination and distribution of work in machinery in general is performed —such as the operation of the moving mass in our present propeller-engines, which constitutes a very important item in the success of the machinery; how the work required to transport

a given cargo a given distance, in different
forms of ships with different speeds is achieved;
all this, yet remains a mere conjecture, is spoken
of as a craft to be acquired only by experience,
and the rationale of the problem has never been
given.

I proposed to the Engineer-in-Chief of the
Navy, Mr. Isherwood, to clear up this subject
of dynamics for the Naval Engineer Cadets,
but the proposition was in vain.

Whatever I proposed in that quarter, whether
based upon true scientific principles which
could not be disputed, or upon ideas which are
avowedly in successful operation in different
parts of the world, was obdurately declined
and invariably overwhelmed with quack reasoning, informing me that what they already did
was perfection, and that every possible idea was
exhausted for ages to come.

The disposition to suppose that we have
reached perfection actually bars the path of
progress in the Navy Department, and that
illustrious Chief, with all his talent, will never
progress, until he finds out that he is behind
the time.

By reason of a want of a proper development
of the science of dynamics, even Mr. Isherwood
has committed serious blunders in his elaborate

works on steam-engineering; a reference to one instance of which will be sufficient to justify this statement.

In the *Engineer's Precedent,* vol. 2, page 26, Mr. Isherwood divides the equivalent of horse-power 33,000 foot-pounds by Joule's dynamic equivalent of heat 772 foot-pounds; and he calls the quotient 42.7461 pounds of water raised one degree Fah., the thermal equivalent of an indicated horse-power! This is nonsense.

The equivalent of one horse-power is a force of 33,000 pounds, moving with a *velocity* of one foot per minute, or the product of *force* and *velocity;* whilst the dynamic equivalent of heat is a force of 772 pounds moved through a *space* of one foot. But space is the product of time and velocity, for which the dynamic equivalent of heat, will be the product of the three simple elements, *force, velocity,* and *time,* which is *work*.

Therefore, if *power* is divided by work, the quotient will be the reciprocal of time, instead of the thermal equivalent of indicated horse-power, as erroneously asserted by Mr. Isherwood.

This error is carried through his two volumes of *Experimental Reseaches in Steam-Engineering,* and he has based thereon some very important calculations **and** decisions, on the efficiency of **different** kinds **of** coal, on **the** evaporative efficiency of different kinds of boilers, and on the economy of the expansion of steam !

Other engineers have thus been led into the same errors, some of which have been repeated in scientific journals.

We have yet no books for **the** schools or **colleges, which** explain the difference between *foot-pounds* **of** *power,* *foot-pounds* **of** *work,* and *foot-pounds* **of** *momentum.*

In order to clear up the science of dynamics, it will be necessary to abolish a number of useless terms which now confuse the subject and to establish **a** specific meaning for the terms retained. To give authority to such a proposition, I submitted a paper on **the** subject **to** the National Academy of Sciences, at its meeting **last January; but** the Academy declined **to** act **upon it, and** informed me that the subject has been sufficiently discussed.

The paper submitted is as follows:—

NAVY DEPARTMENT,
BUREAU OF STEAM-ENGINEERING,
WASHINGTON, Dec. 14, 1865

To the Chairman of the National Academy of Sciences,
Washington, D. C.

SIR: I most respectfully request that you
would invite the attention of your scientific
association, at its next meeting, to the enclosed
papers on the *science* of dynamics, and oblige,

Yours, most respectfully,

JOHN W. NYSTROM,
Engineer U. S. N.

COMMUNICATION TO THE NATIONAL ACADEMY OF SCIENCES, ON DYNAMICS.

The science of dynamics seems yet to be in
an unsettled condition, since students from
different colleges and even from the same college, are found to differ as regards the true
meaning of dynamical terms; and our school-
books seem to be unnecessarily ambiguous on
that subject.

Independently of the numerous terms differently applied, the substance of the subject is
often misconceived, and not altogether rightly

represented. This want of order and perspi-
cuity in the subject is not at all due to intrinsic
causes, and it would seem that the science of
dynamics can be represented in a very clear
and simple form. An effort to this effect is
contained in the accompanying papers.

ON THE ELEMENTS OF DYNAMICS.

FORCE is a mutual tendency of bodies to at-
tract or repel each other. Its physical constitution is not yet known. We only know its
action, which is recognized as pressure and
measured by weight. The unit of weight being
assumed from the attraction of the earth upon
a determined volume of any specific substance;
for example, the force of attraction between
the earth and 27.7 cubic inches of distilled
water, at the temperature of 39.8° Fah., in an
atmosphere balancing 30 inches of mercury,
at the level of the sea—is called one pound
avoirdupois.

Force is the first element of power and work,
and may be likened to length, which is a pri-
mary element in geometry. Force will here
be denoted by the letter F, expressed in
pounds.

VELOCITY is the second element of power and work, and may be likened to breadth in geometry. It is that continued change of position recognized as motion, and is here denoted by the letter V, expressed in feet per second. Velocity is a simple element, although it appears to be dependent on time and space, but the space is divided by the time, and therefore both eliminated from the velocity.

TIME is the third element of work, and may be likened to thickness in geometry. It implies a continuous action recognized as duration. Time is here denoted by the letter T, expressed in seconds.

POWER is a function of the first two elements —force F, and velocity V,—as area in geometry is a function of length and breadth. Power is here denoted by $P = F\,V$, which means that the power P is the product of the force F multiplied by the velocity V. The power so obtained is expressed in foot-pounds, and called dynamic effect, of which there are 550 in a horse-power; or, if the velocity is measured in feet per minute, there will be 33,000 foot-pounds in a horse-power. Power is independent of space and time, but it has often been confounded with work, which essentially depends on time and space.

9

SPACE is a function of the second and third elements—velocity V, and time T,—and may be likened to a cross-section of a solid, which is a function of breadth and thickness. Space is here denoted by $S = V T$, which means that the space S is the product of the velocity V and the time T, expressed in linear feet.

WORK is a function of the three elements— force F, velocity V, and time T. It may be likened to a solid in geometry which has the three dimensions, length, breadth, and thickness. Work is here denoted by $K = F V T$, which means that the work K is the product obtained by multiplying together the three elements—force F, velocity V, and time T.

Work may be denoted by $K = F S$, or the product of the force F multiplied by the space S, where it appears as if the work was independent of time, but the time is included in the space $S = V T$.

Work may also be denoted by $K = P T$, which means the power P multiplied by the time T. Either of the three cases expresses the work in foot-pounds.

Force, velocity, and time are simple physical elements.

Power, space, and work are functions or products of those elements.

It appears that $F\ V\ T$ is the mathematical definition of a trinity of physical elements which governs the material universe. All action of whatever kind, whether mechanical, chemical, or derived from light, heat, electricity, or magnetism; all that has been, and is to be done or undone, is comprehended by this triume function, $F\ V\ T$. It is omnipotent, ubiquitous, and eternal.

I am, at present, stationed in the Bureau of Steam-Engineering of the U. S. Navy Department, where occasions have often arisen to discuss the subject of dynamics with naval engineers, some of whom have studied *Moseley's*, *Bartlett's*, or *Weisebach's Mechanics;* and yet most of them not only differ with me, but also disagree amongst themselves regarding the precise meaning of dynamical terms. They all seem to agree that *time* is included in *power*, but that *time* is not included in work. Their argument runs thus: "The unit of *power* is "33,000 lbs., lifted one foot per *minute*, whilst · "the unit of *work* is one pound lifted one foot, "independent of *time*." Some of them do not recognize the term *power*, and say that *power* is *work* done in a certain *time*.

My own argument is, that space is the product of time and velocity; and when we say per minute or per any unit of time in the ex-

pression of horse-power, meaning a force of so
many pounds raised so many feet in a given
time, we divide the space by the time, and the
result is only force and velocity, the time being
eliminated, as appears in the formulas following.

The popular expression of power is—

$$P = \frac{F\,S}{T}$$

But when $\qquad S = V\,T$

we have $\qquad P = \frac{F\,V\,T}{T} = F \cdot V,$

or T disappearing, and power thus contains no
time.

Work is generally known to be the operation of raising a given weight to a given
height, as a force passing through a given
space. But this height or space cannot be
attained without time and velocity—its consti-
tuent elements—as before stated.

The popular conception of work is

$$K = F\,S,$$

but when

$$S = V\,T,$$

we have

$$K = F\,V\,T,$$

or *time* is one element in *work*.

I am the author of a pocket-book of mechanics
and engineering, of which a copy accompanies

this paper. This book is now to be found in most parts of the civilized world. It was compiled when I was a boy, and as soon as time and means will allow, it is my intention to rearrange the whole book, adding much valuable matter which has not heretofore appeared in print, and putting the subject of dynamics into some kind of standard form. But before so doing, I wish to consult the National Academy of Sciences as to what dynamical terms are to be accepted as proper.

The first question stands thus—

Elements?		*Functions?*	
Force	$= F?$	Power	$P = F V?$
Velocity	$= V?$	Space	$S = V T?$
Time	$= T?$	Work	$K = F V T?$

Is it right to consider F, V, and T as elements?

Is it right to denominate P, S, and K functions?

Are power P, space S, and work K, composed, as indicated by the above formulas?

In chemistry the combination of simple elements is called "a compound." By what corresponding term can we denominate the combination of physical elements?

In mathematics a function is called any

9*

quantity obtained by whatever process or operation indicated by a formula. Accordingly a simple element might be called a function, although it is self-evident that a simple element by itself cannot be considered a function.

When $V = \dfrac{S}{T}$, can we not say that the function S is resolved into its elements?

Let a constant force F, act on a body of weight W, in the direction of the arrow. F and W being measured by the same unit of weight, and no other force acting on W; then we know that the acceleratrix of the force F will be—

$$G = \frac{g\,F}{W}$$

when $g = 32.166$, the acceleratrix of the force of gravity at the surface of the earth.

Let the force F act for any length of time T, then we know from the law of gravity, that the velocity attained will be—

$$V = G\,T.$$

Let v denote the velocity at any time t, then the power in operation will be—

$$P = F v,$$

and the differential work will be

$$dK = F v \, dt.$$

but

$$v = G t$$

and

$$dK = F G t \, dt,$$

the work

$$K = \int F G t \, dt = \frac{F G t^2}{2}.$$

Let the work be integrated from $t = o'$ to $t = T$ we have—

$$K = \frac{F G T^2}{2}, \qquad \cdots \qquad 1$$

$$K = \frac{F V T}{2}, \qquad \cdots \qquad 2$$

and

$$K = \frac{F V^2}{2 G}, \qquad \cdots \qquad 3$$

which are the three forms of work accomplished by the force F acting on a body W.

We know that $F = \dfrac{G W}{g}$, which inserted in the above formulas, will give the same work, expressed by the weight of the moving body.

$$K = \frac{W G^2 T^2}{g}, \qquad \cdots \qquad 4$$

$$K = \frac{W G V T}{2 g}, \qquad \cdots \qquad 5$$

$$K = \frac{W V^2}{2 g}, \qquad \cdots \qquad 6$$

These formulas give the three forms of work concentrated in a moving body W. We have thus six different formulas expressing the same work K. Cannot this work be denominated by one generic term?

The case is the same for all circular motion, where $V = \dfrac{2 \pi r n}{60}$; $r =$ radius of gyration, $n =$ revolutions per minute.

Vis-viva, or living force, expressed by $M V^2$, seems to be the most unfortunate term in dynamics, as it has caused so much controversy and confusion. When M means the mass of a moving body, the term $M V^2$ represents double the work concentrated therein, the true work being that represented by formula six. *Vis-viva* is, therefore, substantially the same as work. In fact, if *vis-viva* means *living force,* there is no more *vis-viva* in a moving mass than in one at rest, and therefore it does not express what it means. It requires $F V T$ and nothing else to set a body in motion. It requires $F V T$ and nothing else to bring a moving body to rest. It is $F V T$ and nothing else that can change the motion of a body. F being an external force, equal to the force of inertia in the moving body. I would, therefore, propose to reject the term *vis-viva* in its present acceptation in dynamics.

In the estimate of foot-pounds of power, I have made a deviation from **Watt's rule.** The unit 33,000 pounds raised one foot per minute I think is very unnatural. A velocity of only one foot per minute cannot be clearly conceived; it is only 0.2 or ⅕ of an inch per second, the velocity of a snail; on the other hand the weight of 33,000 pounds, or about 15 tons, is too large, and very few can form a clear conception of its magnitude. A horse cannot lift directly a weight of 15 tons; the ordinary pull of a horse is 200 to 300 pounds, and a horse cart-load is about one ton. Therefore it is my humble opinion that the foot-pounds of power ought to be brought nearer the ordinary performance of a horse, or one pound lifted one foot per second as a foot-pound, of which there will be 550 foot-pounds per horse power. A velocity of one foot per second is conceivable, and 550 pounds can be lifted by a horse. This kind of foot-pounds is used in most parts of Europe. The **Swedish** horse-power is 600, and the German 513 foot-pounds.

Would it not, therefore, for the reason stated, be better to call a horse-power 550 foot-pounds, instead of 33,000? Would it not be well, also, to establish an additional unit for work?   One pound raised a space of one foot is called one unit of work, by which the labor a man is capable of performing in a day, will be represented in millions.   The power of an ordinary man is about fifty pounds raised with a velocity of one foot per second, which will amount to between one and two millions of foot-pounds of work in one day.

Let us assume the work accomplished by one horse-power in a time of one hour to be a unit for physical work; which will be the same as that of eleven men working one hour, or that of one man working one day of eleven hours. In order to clearly distinguish this unit from that of power, let it be called a *workmanday*, which means a man's day's work.

A *workmanday* expressed in force and space will be $550 \times 3600 = 1,980,000$ foot-pounds.

All kinds of work can be estimated in workmandays, such as building a house, steamboat, or a bridge, digging a canal, ploughing the ground, steam-boiler and gunpowder explosions, the capacity of heavy ordnance, &c. &c.

Would it be proper, then, to introduce such unit as this for *work?*

The National Academy of Sciences declined to answer the foregoing questions at the time, and as I do not feel disposed to lay the subject aside, have now published the same in my Pocket-Book of Mechanics and Engineering, and hope the Academy of Sciences, at its convenience, will give me the benefit of its criticisms.

The attention of our scientific men seems to be wholly absorbed in the polarized light, the compound microscope, and spectrum analysis; which indeed are very interesting subjects, and we hope may ultimately lead to the revelation of the physical constitution of light, heat, and force, and thereby relieve us from our present method of generating power by means of the cumbrous steam-boiler. But they have gone so far and deeply into the sciences that they have left us practical engineers far behind, toiling in the dark, and when we hail them we receive no answer.

If we could only succeed in bringing steam-engineering under the compound microscope, or if it would produce a line in the spectrum, it might be profoundly analyzed by our scientific professors. But although our naval engineering was magnified some thirty millions ($?) in the case of the light draft monitors, a certain Fox was not detected in it, whilst in the spectrum he could not possible have escaped making a line.

In a period when steam-engineering is not considered sufficiently important, or is not sufficiently advanced, for the Corps of Naval Engineers to be worthy to be represented in the National Academy of Sciences, we cannot expect the members of such an Academy to be familiar with, or to appreciate what is wanted in the practical operation of machinery, with which they have no connection.

The science of dynamics is represented in its simple form in the tenth edition of Nystrom's Pocket Book, but the space is there so crowded that it does not admit of full explanation with illustrations.

About a year ago, when I told Captain Fox that I had some valuable matter on hand which would be useful for the naval engineer cadets, he answered, "If you have anything "new, you just take out a patent for it; we have "a patent office for that purpose."

Captain Fox then also told me about what Democratic Governments do not do, and what Monarchical Governments *do* do, very much in the style of his previous observation.

I wish Captain Fox to know that I believe myself to have a better opinion of a Democratic or Republican form of Government than he has. My humble proposition had nothing whatever to do with the form of Government. Captain

Fox, however, left the impression on me that his ideas about Monarchical Governments are what he has learned in Shakspeare, and as it was in Europe a hundred years ago.

The Navy Department is now attempting to reorganize the Corps of Naval Engineers. Captain Fox tries his best to subordinate the engineers to the Line Officers, and the engineers in the Department strain their efforts to secure rank and position, all with self-interest in view, but no proposition seems to be offered to prepare the new corps of engineers (by receiving the proper learning) to maintain with dignity whatever rank and position may be assigned to it. Give the corps a thorough technological education, and it will become able to take care of itself respecting rank and position; for in the present feeble conception of the value and importance of mechanical skill and of the range of knowledge in steam-engineering prevalent in the Navy Department, it is useless to argue about questions of rank, position, and responsibility of the corps of engineers.

The Washington Navy Yard contains a very extensive mechanical establishment for the building of marine engines and boilers, the superintendence of which is now intrusted to a mechanic brought up in that place, and with

10

no further scientific education. However accomplished this mechanic might be in his limited profession, he cannot possibly fulfil the requirements of such an establishment in connection with the different mechanic arts and sciences which it involves; neither can he command the respect requisite in the proper execution of such a responsible office.

These remarks were made in the Navy Department, and were answered thus: "All the drawings are made here in minute details, and there is no more knowledge necessary in the yard than to follow the drawings."

Such expressions of disregard for the knowledge required in the execution of work were frequently met with.

The Navy Department, where the drawings are made, is some three miles from the yard where the work is executed. The chief draughtsman visits the works perhaps once a week, and remains there an hour or two, which is considered a sufficient connection between the drawing-room and the workshops. The absurdity of this arrangement can well be conceived, for we know by experience, in well-regulated establishments, that for the proper execution of the drawings, in regard to economy and utility in the work, the draughtsman requires constant access to the patterns, pattern-shop, and to the different branches of the establishment; and that there is constant consultation going on between the draughtsmen and the foremen in these different branches.

In Washington these parties are separated by a distance of several miles, in consequence of which the character of the design, the economy and the progress of the work, suffer considerably. However able or talented draughtsmen or engineers may be when entering the Navy Department, they will soon be spoiled, which is readily evinced in their fancy design of machinery with mouldings and ornaments, not to be found nowadays outside the Bureau of Steam Engineering.

These draughtsmen, I believe, are all Engineers in the Corps, and seem to display a goodly array of talent. Many of them have received collegiate educations, are accomplished mathematicians, and well versed in the physical sciences, but for want of a technological education, they are naturally deficient in *the application of their scientific knowledge;* and much of what they would be able to apply and cultivate, cannot be utilized when thus separated from

the field where the seed of their education
ought to be planted.

The Navy Department is now about to pump
steam-engineering into Line Officers at the Naval Academy, Annapolis, and to make engine-
drivers (as they call it) out of ensigns, masters,
and midshipmen, by sending them to sea, and
having them stand watch in the engine-room;
by which means it is expected to make steam-
engineers in the space of two years! How easy
the Navy Department must think steam-engineering to be! This mode of making steam-
engineers is surely the greatest invention of
the age; and Captain Fox, the ostensible author of it, had better take out a *patent* for that
bubble before it bursts; "we have a patent
office for that purpose."

During the war Captain Fox kept in his
room models of machinery and vessels submitted by civilians to the Navy Department.
He decided what was to be done and not to be
done in questions of naval engineering. He
undertook to superintend the construction of
vessels in the Bureau of Construction. He
selected models for the constructors to make
drawings of.

Now, for the sake of argument, let us suppose that Captain Fox is, by nature, gifted with

a peculiar faculty which enables him to guess which model of vessel is best for a desired purpose. A vessel is built from that model, and proves to be satisfactory or not. Neither Captain Fox nor the constructors in the navy are yet able to judge or record the peculiarities of that vessel, in form of scientific arguments, why it did or did not come up to what was anticipated. In case it proved to be a good vessel, there might still never be another one built like it, and Captain Fox's superintendence will thus only satisfy his own personal ambition, without leaving recorded and permanent results for the future benefit of the navy.

Now if Captain Fox had allowed the introduction of the science of shipbuilding into the Department, the achievements of his own talent might have been recorded and perpetuated for the benefit of the country.

In my humble judgment, I believe Mr. Lenthal has attained great perfection in the construction of ships, for which reason I was very anxious to give his lines a scientific investigation, but he would not allow me to see his best drawings. The ship's drawings intrusted to me in the Department, through the request of Mr. Isherwood, for information required in the Bureau of Steam-Engineering, were of some

twelve different vessels, of which only two were
of Mr. Lenthal's construction. It appeared that
Mr. Lenthal would not allow me to explore his
late ships' drawings, even at the request of
Mr. Isherwood. The engineers in the Department are not allowed to see the ship's drawings, except through the kindness of some
clerks, who, in Mr. Lenthal's absence, ran the
risk of letting some of them be seen. I did
not avail myself of such an opportunity, for
although I have the highest regard for Mr.
Lenthal as a constructor and shipbuilder, his
drawings would not warrant such a proceeding
on my part.

The engineers in the Department generally
evince a strong appetite for learning, and when
I received some few old ships' drawings from
the Bureau of Construction, they generally
surrounded them with a manifest anxiety to
gather information, and remarked that "it
"must have broken Mr. Lenthal's heart to have
"given these drawings out of his office."

In case it were the policy of the Government
to keep their ships' drawings secret from a desire to promote the interest of the country, I
would heartily acquiesce; but it is a question
of considerable importance whether such a

policy would not act detrimentally rather than otherwise.

If a knowledge of shipbuilding is to be restricted only to a chosen few constructors in the navy, it would be by a rare accident only that those of adequate capacity would be hit upon, while if thrown generally open, it would scarcely fail to reach many whose talent in that particular would be unquestionable.

In reality, this attempt at secrecy is a folly, for in view of the conspicuous enterprise and fertility of the American mind, we need not fear to be behind the time by a liberal diffusion of useful knowledge, for unless we sow, there can be no harvest.

The farther we look back on the art of shipbuilding, the more secret it appears to have been kept, until, at the present time, some of our first shipbuilders not only freely expose their drawings to observation, but even allow them to be published.

The lines of the fastest and best of the European steamers, namely, the "Persia," and others, are published in McKensy's Shipbuilding, and those of the "Scotia" also are published in Scott Russel's work.

The sloop of war of the *Wampanoag* class, intended for great speed, we have reason to suppose, have the most perfect lines of the day, which, in connection with their intended great propelling power, afford a very interesting field for scientific investigation, the result of which might be of great value to the Corps of Engineers, but will now probably be lost to them through a personal jealousy.

This jealousy is by no means limited to the Navy Department, but is met with in all directions, and frequently intercepts scientific inquiry. It is an epidemic disease which can be cured only by technological education.

Once, in a scientific meeting, efforts to explain the science of steamboiler explosions, and how to prevent the same, were silenced by the president of the meeting calling me to the chair and whispering, "Don't say anything about "boiler explosions." The discussion was accordingly dropped and lost.

An explanation of the cause of steamboiler explosions is a question demonstrably within the reach of science, as much so as a problem of geometry.

In many cases explosions are indicated a long time before they occur, and could be easily prevented. The terrible explosion on board the steamer Sultana, on the Mississippi, and a great many other similar disasters, were indicated several hours before they occurred, all of which could have been prevented, and a great number of lives saved.

ON STEAM-BOILER EXPLOSIONS.

It has hereinbefore been explained what is meant by *work*, namely, the product of the three simple physical elements *force F*, *velocity V*, and *time T*, or work

$$K = FVT. \qquad . \qquad . \qquad . \qquad 1$$

The heat required to elevate the temperature of one pound of water one degree Fah. is assumed to be *one unit of heat*, and found to be equivalent to the dynamic effect of 772 foot-pounds; *or one unit of heat* can raise 772 pounds one foot, or one **pound to a** height of 772 feet.

The heat in the steam **and** water in a steam-boiler is equivalent to such a proportion of work, or the steam-boiler is a reservoir of work which is generally dealt out in *homœopathic* doses to work a steam-engine. But when the entire stored work, $K = FVT$, is suddenly started, as in the case of the bursting of a boiler, the steam and water in the form of a foam, impelled by the heat, performs a proportionate destruction in the explosion, the force of which will be

$$F = \frac{K}{VT} \qquad . \qquad . \qquad 2$$

From this formula we see that the less time occupied by the explosion the greater will the force be, or if the time is infinitely small, the force of the explosion will be infinitely great.

It has been assumed that explosive gases are sometimes formed in steam-boilers, which cause explosions; but the concentrated work in the steam and water is amply sufficient to perform the destruction without the aid of any further explosive gas.

DESTRUCTIVE WORK OF STEAM-BOILER EXPLOSION.

When steam-boiler explosions take place, the inclosed water is resolved into one volume of boiling hot water, and one volume of steam, as follows :—

Notation of Letters Prior to Explosion.

W' = total weight in pounds of the water in the boiler under full steam pressure.

w' = pounds of water evaporated in the explosion.

h = units of heat per pound in the water W'.

H = units of heat per pound in the steam of pressure P.

H' = units of heat per cubic foot in the steam P.

P = pressure of steam in pounds per square inch.

V = volume coefficient of steam.

Then the water evaporated in the explosion will be

$$w' = \frac{W' H (h-180.9)}{824.8}. \qquad . \qquad . \qquad . \quad 3$$

The destructive work K of the explosion will be in foot-pounds.

$$K = \frac{W' H P}{5.728 \, H' \, V}(h-180.2)(V-1)(2.3 \, log. P$$
$$-1.6848298) \qquad . \qquad . \qquad . \quad 4$$

By exemplifying this formula, it will be found what an enormous destructive energy there exists in steam-boilers.

For values of the letters, and also examples, see Nystrom's Pocket-Book, tenth edition.

When the steam pressure in any part of a boiler is suddenly removed, the entire work concentrated therein is started with a violence proportioned to the removed pressure, and the steam and water, in the form of a foam, strike the sides of the boiler, by which the work is suddenly arrested. If the time of arresting the work is infinitely small, we see from the formula 2, that the force of the work will be infinitely great, and thus the boiler explodes.

The sudden removal of pressure, which invariably leads to the explosion, is derived either from bursting or by collapse of some part of the steam-boiler; for instance:—

1st. By long use boilers become corroded and give way in some unexpected place, which ought to have been detected by inspection, or in cleaning.

2d. The general construction with staying and bracing of steam-boilers is often very carelessly executed and results in explosion. This kind of explosion is often indicated, long before the accident occurs, by leakage of the boiler; when the engineer, not suspecting the approaching danger, limits his remedies generally at efforts to stop the leak. The leakage from bad calking or packing is easily distinguished from that of bad or insufficient bracing. In the latter case the fire ought to be hauled out, the steam blown off very slowly and carefully,

so as to make as little disturbance in the work
as possible, or it would be safest to work off
most of the steam by the engine; after which
the boiler should be secured by proper bracing.*

* TERRIBLE STEAMBOAT DISASTER.—Memphis, March 5.
The *R. J. Lockwood*, bound from New Orleans to St. Louis,
exploded, about seven o'clock last evening, while eighteen
miles below this city, and afterwards burned until she
sunk. She was inspected at New Orleans on last Wednes-
day, and pronounced seaworthy. After running a day or
two her boilers were discovered to be in a leaky condition.
Captain Howard proposed to the engineer to stop at Helena
and repair, but the engineer thought it unnecessary until
the boat should arrive at Memphis. Thus the delay
proved fatal. The explosion tore away the cabin as far
back as the centre, killing twenty persons instantly, and
scalding, wounding, or otherwise injuring about twenty-
five others. Fortunately the *M. S. Mepham* was coming
down at the time, and rendered most timely aid to the
distressed passengers and officers of the boat. As the
Lockwood caught fire immediately, the *Mepham* rounded
to and landed her bow against the stern of the ill-fated
steamer, thus saving every person not killed by the explosion.

A number of cabin passengers, crew, cooks, and negro
firemen were lost; but their names are not known. All
the lady passengers, besides the two chambermaids, were
saved. I think the number of killed amounts to forty
or fifty, as survivors state there were a number of deck
hands and deck passengers on the lower deck, who were
killed by the explosion, and whose names are unknown.
—*St. Louis Republican*.

11

3d. Explosions are sometimes caused from low water in the boiler, but more rarely than generally supposed. When the fire-crown and tubes are subjected to a strong heat and not covered with water, the steam does not absorb the heat fast enough to prevent the iron from becoming too hot so that it cannot withstand the pressure, but softens and collapses. Sometimes, when the boiler bursts, the tubes and flues may also collapse by the force of explosion, when it has been erroneously supposed that the explosion was caused from such collapse.

4th. Steam-boilers often burst by strain in uneven expansion or shrinkage, occasioned by the fire being too quickly lighted or extinguished. Explosions of this kind frequently occur on Saturday nights or Monday mornings, or before or after a holiday. The reason for this is, that on Saturday nights the engineer generally puts out the fire too quickly in his haste to go home, by which the most heated part of the boiler is too suddenly

Cases of this kind are occurring over and over again. It is perfectly clear that the engineer on the steamer *Lockwood* did not understand the character of the leak, and he was most probably unfamiliar with the rudiments of the operating natural principles involved in the subject.

cooled, and may burst in too rapidly shrinking. On Monday mornings the engineer may be late, and in his hurry to get up steam in time, throws in dry shavings or wood which heats the flues or tubes in the boiler too rapidly whilst other parts remains cool, when the unequal expansion thus occasioned may strain some parts sufficiently to burst, and explosion follows.

Throughout the week the fire is not hauled out but tossed against the bridge and fresh coal thrown on the top of it in the evenings, where it will remain and keep the boiler hot until started for work the next morning.

5th. It is a very bad practice to make boiler ends of cast iron, composed of a flat disk of from two to three inches thick, with a flange of from one to two inches thick with cast rivet-holes. The first shrinkage in the cooling of such a plate causes a great strain, which is increased by riveting the boiler to it. Any sudden change of temperature, therefore, either in starting or putting out the fire, might crack the plate and thus occasion an explosion. Such accident may be avoided by making the cast-iron ends concave and of even thickness.

6th. In cold weather, when the boilers have been at rest for some time, they may be frozen

full of ice, then when the fire is made in them
some parts are suddenly heated and expand,
whilst other parts still remain cold, causing an
undue strain, which may also burst the boilers.
Such accident can be avoided by a slow and
cautious firing.

7th. Sometimes a great many boilers are
joined together by solid connections of cast-iron
steam-pipes, which expand when heated, whilst
the masonry inclosing the boilers contracts.
Should such a steam-pipe burst from expansion
or shrinkage, explosion will likely follow in all
the connected boilers, of which numerous examples have occurred. Such accident may be
avoided by making the connections elastic, or
free to expand or contract without straining the
boilers. The fragments of one exploded boiler
striking the next, also cause continued explosions of several boilers.*

Steam-boiler explosions are thus not always
caused by pressure of steam alone, but often by
expansion and contraction of the materials of
the boiler. A boiler which is perfectly safe
with a working pressure of 200 pounds may
explode with a pressure of only 20 pounds to
the square inch.

---

* Such a case occurred lately at Harrisburg, Penn.,
where eight boilers exploded in one succession.

Four hundred and ninety lives were lost by boiler explosions in this country from the fifth of January to the ninth of February, 1866.

If the president who stopped my discussion on steam-boiler explosion, were seated on the top of a boiler when it exploded, and were blown up a few hundred feet in the air, and came down comfortably and unhurt, it would be interesting to ascertain whether he would stop a discussion on steamboiler explosion at the next meeting, or if he would not like to know how it happened ?

It will perhaps be remarked that it is very improper to take so much upon myself, and speak so plainly about all these things, as has been done in this book, but I have been kept so long in silence, whilst the neglect and ignorance which still control the subject have consigned so many thousands of souls to heaven, and so many millions of dollars in the opposite direction, that these protestations cannot be suppressed.

The appropriate remedy for the evil is THE ESTABLISHMENT OF TECHNOLOGICAL INSTITUTIONS THROUGHOUT THE LAND.

11*

A REVIEW

ON THE

SUBJECT OF SCREW PROPELLERS.

The screw propeller has at length become such a familiar instrument, that every engineer knows how to construct one. It is frequently declared to have reached perfection, and it has even been said that "sufficient has been written to put this subject in a true light;" nevertheless, the writer has no apology to offer for bringing forward the following views on this hackneyed theme.

The common straight-bladed screw will first be considered; afterwards the same with expanding pitch; and then the different forms of curve-bladed screws in like manner.

It will not be necessary, in this article, to enter into minute details of the construction of the screw, as the same will be found in a practical treatise on the subject now in preparatoin for the press.

Fig. 1

Fig. 2

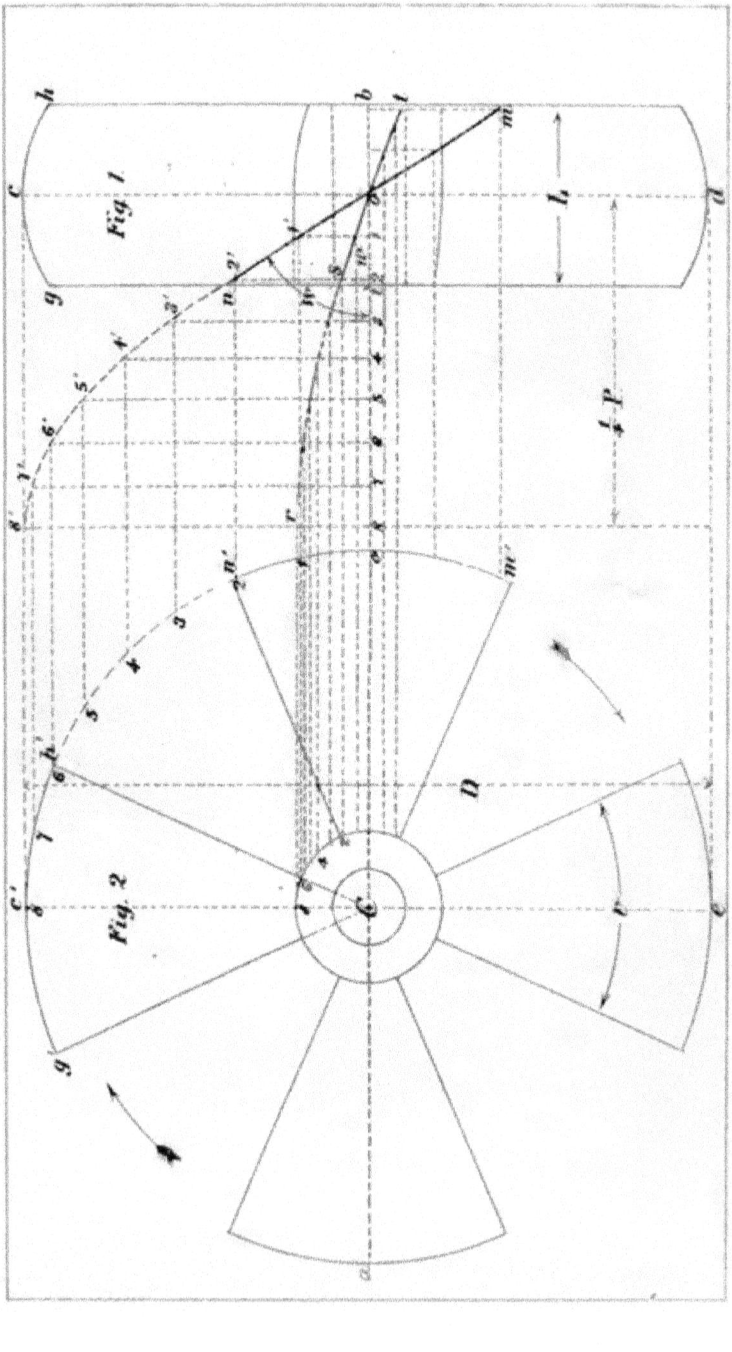

TO CONSTRUCT A PLAIN SCREW PROPELLER WITH A UNIFORM PITCH. (PLATE II.)

Draw the line $a b$ (Plate II.), and the lines $c d$ and $e c'$ at right angles thereto. Draw the circumference of the propeller, fig. 2, with the given diameter D; divide the quadrant $o' c'$ into any number (say eight) equal parts, and number them as shown in the figure. Set off from o, fig. 1, one-quarter of the assumed pitch P, and divide it into an equal number of parts with that of the quadrant $o' c'$. Draw from each division point the rectangular ordinates $8'$, $7'$, $6'$, $5'$, $4'$, &c. &c., and the intersection of those by lines of equal numbers will constitute points in the helix of the screw.

Repeat the same operation with the hub, and draw the two helixes $8' n o m$, and $r s o t$, as shown in the figure. Set off the assumed length L, fig. 1, then project $n o m$, to $n' o' m'$, fig. 2; join n' and m' with the centre C, which thus forms one blade of the propeller. Complete the assumed number of blades (say four), then the curve or helix $g' c' h'$, fig. 2, projected on fig. 1, will appear as $g c h$, so that $c g = c h = 8' 7' 6'$.

Let v denote the projecting angle of each blade, then—

$$P : L = 360 : V.$$

Pitch $$P = \frac{360\,L}{V},$$.   .   .   1

The angle of the blades with the axis of the propeller, at the periphery, will be—

$$\text{tang. } W = \frac{\pi\,D}{P},$$.   .   2

and—

Pitch $$P = \frac{\pi\,D}{\text{tang. } W},$$.   .   .   3

TO CONSTRUCT A PROPELLER WITH A COMPOUND EXPANDING PITCH. (PLATE III.)

The pitch of a propeller may expand in the direction of the axis as well as in the direction of the radius or generatrix. Let the generatrix have a uniform motion around the axis, and an accelerated motion in the direction of the axis; then the pitch will expand in the direction of the axis. When the generatrix has a quicker motion in the direction of the axis at the periphery than at the axis, then the pitch will be expanding in the direction of the radius or generatrix, and when the pitch is composed of the two expansions, it may be denominated a compound expanding pitch.

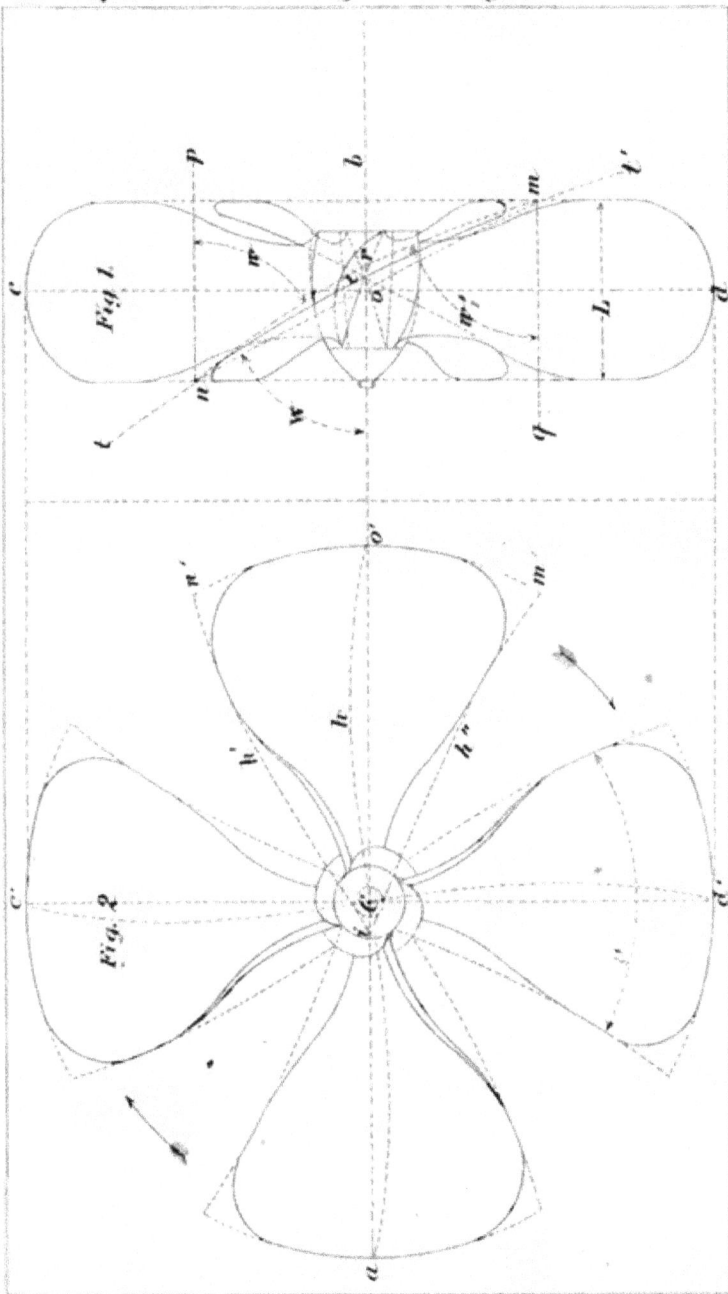

Fig. 1.

Fig. 2.

The propelling energy of the blades near the hub of the propeller is very inconsiderable, and acts mainly to agitate the water, for which reason it has been proposed to construct propellers, so that *the pitch at the periphery is to the pitch at the hub, as the velocity of the pitch of the periphery is to the velocity of the ship:* so that the water will pass through at the hub without being further disturbed. Having given, say the diameter, $D = 12$ ft.; length, $L = 3$ ft.; mean pitch, $P = 18$ ft.; it is proposed to construct a propeller with a compound expanding pitch. (Plate III.)

Draw the centre lines $a\,b$, $c\,d$, and $c'\,d'$. Calculate from formula 2, the angle

$$\text{tang.}\ W = \frac{3.14 \times 12}{18} = 2.0933 = \text{tang. } 69° \ 30',$$

the angle required.

Draw the arc W, fig. 1, and set off the angle $W = 64° \ 30' = n\ o\ o'$; draw the dotted line $n\,o\,m$. Set off the length $L = 3$ feet. Draw from n and m, the lines $n\,p$ and $m\,q$, parallel to $a\,b$. Assume the pitch to expand from 16 feet at m, to 20 feet at n, then calculate the angles w and w' from formula 2.

$$\text{tang.}w' = \frac{3.14 \times 12}{16} = 2.356 = \text{tang. } 67°$$

$$\text{tang.} w = \frac{3.14 \times 12}{20} = 1.884 = \text{tang.} 62°$$

Set off the angles w and w', as shown in fig. 1; draw the dotted lines $t\ r$ and $r\ t'$. Draw a curved line $n\ e\ m$, which will tangent $t\ r$ at n, and $t'\ r$ at m, then $n\ e\ m$ forms the outer edge of the propeller, bladed with the required expanding pitch.

Draw the circumference of the propeller fig. 2, project n to n' aud m to m'. Draw the curved generatrix about as $o'\ h\ C$. Draw from n' and m' the genatrixes h' and h''—of the same curve as h—so that they meet on the other side of C somewhere at i; then the helicoidal surface C $h'\ n'\ o'\ m'\ h''$ forms the blade of the required compound expanding pitch.

A propeller of this construction will not be a true screw, on account of the generatrixes h' and h'' not meeting in the centre C. The error is greatest in the centre or near the hub. But when the blades are fashioned off, as shown on the plate, the error is inappreciable in practice. This kind of propeller is now made to the extent of hundreds or perhaps thousands, by one of the most experienced firms in this country, namely Neafie & Levy, of Philadelphia.

Although the construction is not conducted strictly as here described, their propellers are substantially the same, for on account of their great experience the patterns are got up entirely by "rule of thumb," and generally produce good propellers, which have the additional advantage (so much sought for) of having " no science in them." They have on hand a great number of patterns of different diameters. When, therefore, a screw is ordered, they select the pattern of nearest diameter, and the difference, if any, is cut out or filled in, in the mould.

They rarely ever make a drawing of a screw; but when an order is given for one, the draughtsman gets the required angle W of the blade at the periphery from a diagram constructed for that purpose. The draughtsman writes down the diameter of the propeller, the angle of the blade, the diameter and bore of the hub, on a printed form of card, which is given to the moulder, who then, as before stated, selects the nearest pattern.

The blades on the patterns are generally made loose so that they can be set at the desired angle, which is done by an instrument composed of a spirit-level and a graduated circle arc. The angle and pitch are thus adjusted with greater precision in the foundry, than by

the draughtsman from the diagram. The angle
of the blades at the hub, in such a method, is
not considered of any importance, in relation
to that at the periphery.

THE U. S. NAVY PROPELLER AS CONSTRUCTED BY CHIEF ENGINEER B. F. ISHERWOOD.

Plate IV. represents the propeller constructed
in the U. S. navy department for a great many
years past.

The generatrix for the screw is a circle arc
with the centre c, fig. 1; but the generatrix at
right angle to the axis is a curved line $a\,m\,b$.
Either of those generatrixes will generate the
same helicoidal surface, or if the propeller was
made straight like $a\,m\,b\,f\,n\,d$, fig. 1, the blade
would appear $a\,m\,b\,f\,n\,d$, fig. 2, that is to say,
that if the part $a\,m\,b\,c$ is taken off and put on
the blade at $d\,e\,f\,n$, the same propeller would
appear as the dotted lines, whilst the direction
of the helicoidal surface, and the propelling
efficiency of the propeller, would in both cases
be the same, because the outer configuration of
the blades (as stated by Mr. Isherwood) does
not affect the propelling efficiency.

Mr. Isherwood says in his *Engineer Prece-*

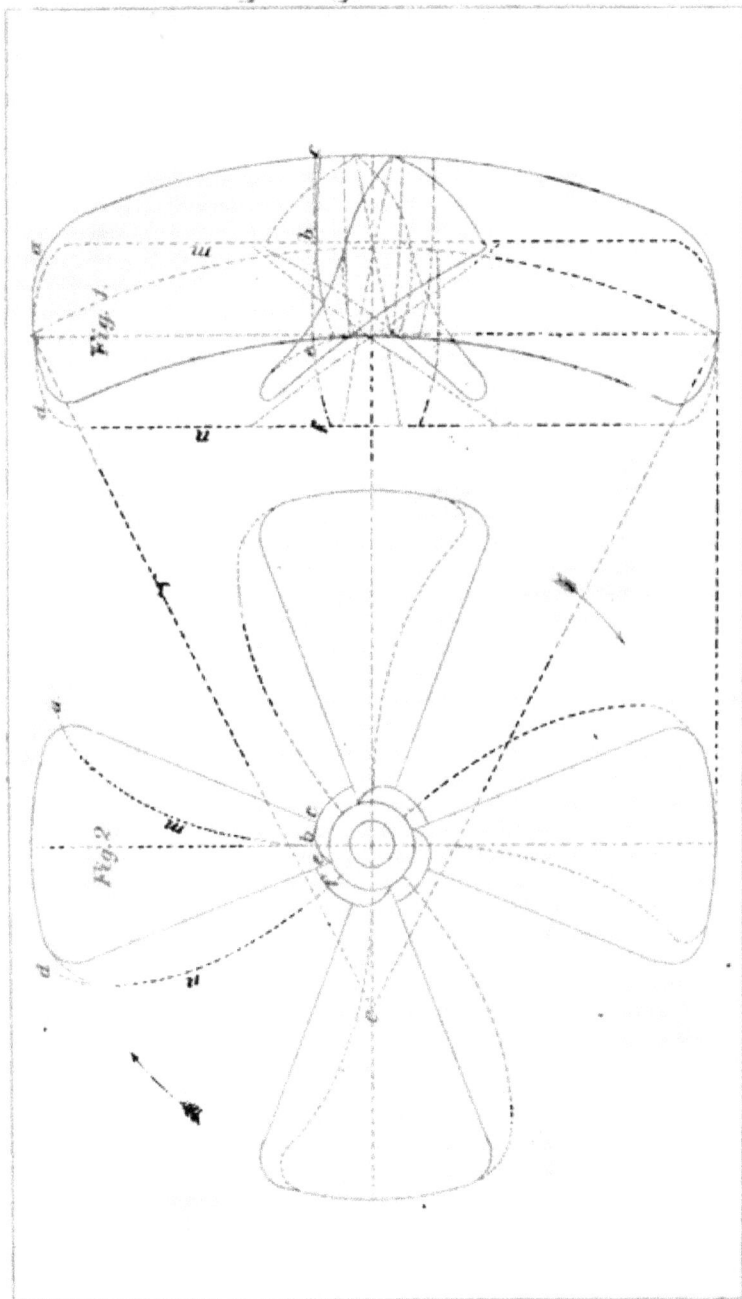

Fig. 1

Fig. 2

dents, vol. i. page 96: "The only improvement possible on the true screw of uniform length from hub to periphery, is that due to the use of an expanding pitch or curved directrix."

I have examined Mr. Isherwood's propeller, which, as it has found a place in the Navy, appears worthy of a few remarks.

The peculiarity of Mr. Isherwood's screw consists in the formation of the helicoidal surface by a curved *generatrix,* instead of a straight *generatrix,* as usually employed.

It may be necessary to define what is meant by a generatrix. Let any right line whatever be taken and called the *axis* of the helicoid or screw, and let any other line curved or straight of definite length be taken, lying at any inclination to the axis, and with one extremity touching the axis; this line is the *generatrix* of the *helicoid* when it is moved simultaneously with a rotary speed around the axis, and a rectilineal speed along the axis.

As previously remarked, Mr. Isherwood uses a curved line for the generatrix, instead of a straight line, as usually used; and the object of using the curved generatrix is (appa-

12

rently) to prevent the alleged loss of effect, caused by the centrifugal force imparted by the rotary motion of the helicoidal surface to the particles of water with which it is in contact. Before proceeding to cure an alleged evil, it may be advisable to ascertain if it exist, and to what extent.

If the helicoid had no slip, and moved through water a distance equal to its pitch per revolution, there evidently could be no centrifugal force communicated to the water in contact with it, for as the helicoid continuously advanced, it would no sooner press any molecule of water than the pressure on that molecule would be removed by the advance of the helicoid. The helicoid in this case would be in contact with the molecule but an infinitely short time, and of course could endow it with no centrifugal motion. But if the helicoid have slip, the effect of that slip is to keep the molecule of water in contact with the helicoid for a time proportionate to the slip, and consequently to endow it with a proportional centrifugal force. If the slip be considerable, the throwing off of the water at the periphery of the helicoid may become sensible, in consequence of that force and the property of water to escape by the easiest road. This effect was observed in the experiments of Taurines, which were performed on *fixed* screws, not advancing rectilineally, but having only a rotary motion on the axis; by which arrangement the slip amounted to unity. The amount of centrifugal force imparted to the particles of water by the rotating helicoidal surface in contact with them was, however, very trifling, even under the most favorable condition of maximum slip;• for the very effect of that slip was to discharge the water from the helicoid, the vacuity being filled by fresh water flowing in. Having thus determined the existence of a small amount of this centrifugal force, and the condition of slip modifying it; let us inquire whether, even in the event of the amount of this force becoming considerable, there would result any loss of effect.

And, first, what would be the nature of that loss of effect, if any exists? It would be shown in the *increased slip of the screw*, for the following reason. The water being thrown off radially in all directions from the axis by the centrifugal force communicated to it by the revolving screw, there would be a vacuum about the axis, provided the centrifugal force forming the vacuum exceeded the force with which the surrounding water would flow in to fill it, and

the resistance to the screw would be decreased
in proportion to the extent of their vacuum;
that is, the slip of the screw would be increased.
*The loss of effect, therefore, due to this centrifugal
force would be measured by the increased slip of
the screw.* But if the water flowed into the
vacuum as fast as it was formed, the resistance
to the screw would evidently remain the same
as though there were no centrifugal force in
action; and this is what actually occurs in
practice.

Let it then be considered the depth to which
the axis of a thirteen feet diameter screw
(mean size) is immersed, and the consequent
pressure of water about it, and then the slip
found in practice ranging from fifteen to thirty
per cent. for maximum, and it will be seen how
enormously the effect of any centrifugal force
must be exaggerated to make it productive of
a vacuum at the axis of the screw.

Supposing now a centrifugal force to be given
to the molecules of water in contact with the
helicoidal rotating surface of any amount less
than requisite to produce a vacuum at the axis;
would it be attended with loss of effect by the
screw? Evidently not, for the following reason :—

With a straight generatrix touching the axis,

the lateral component of the oblique surface of
the screw if tangential to a cylinder, by which
the screw may be supposed to be enveloped;
or is at right angles to the radii of that cylinder.

With a straight generatrix touching an inner
cylinder, having a common axis with the screw,
the lateral composition is no longer at right
angles to the radii of the enveloping cylinder,
but either converges to, or diverges from, the
axis, as either the acute or obtuse angle is
used for propulsion. In both cases the component in the direction of the axis continues
the same; that is, parallel to the axis. Now it
is the component in the direction of the axis
that alone propels; hence the slip of screws of
the same diameter, pitch, and length, but the
one having a straight generatrix touching the
axis, and the other a straight generatrix touching tangentially a cylinder having the same
axis as the screw, should be the same. The
only effect, then, of converging the lateral component to, or diverging it from the axis, is to
cause a flowing of water to or from the axis,
proportioned to the obliquity of the lateral
component to the axis, and as this component
does not affect the propulsion, it is obviously of
no importance what may be its direction. There
can be no loss of labor attending it, for all

12*

angles are in function of form equally efficient for propulsion, consequently their components are equally so.

The same slip, or, in other words, the same resistance, as obtained from the water, with screws of equal diameter, pitch, and length, whether they have straight, inclined, or *curved* generatrices; for the curved generatrix is obviously but a modification of the inclined generatrix; that is, it is an inclined generatrix, whose inclination momentarily changes. It is therefore governed by the same principles.

There is, however, a *practical* disadvantage and loss of labor attending the use of a curved generatrix, though non-theoretically; that is, in function of form; for the friction of the helicoidal surface on the water, which, in differently proportioned screws, varies from 10 to 20 per cent. of the power applied; being as the surfaces with equal speeds, and the surface of a screw of a given diameter, pitch, and length, or fraction of one convolution of the thread, is greater with the curved than with the straight generatrix, because the arc of a circle is greater than its chord; it follows then, that in otherwise similar screws, the greater the curvature of the generatrix, the greater the loss by friction. The only real loss of power attending the imparting of a centrifugal force to the molecules of water is that due to their momentum, that is, will be expressed by multiplying the weight of water, to which centrifugal force is imparted by the square of the speed, with which it flies off radially from the centre. In the most favorable case of maximum slip, that is, a slip of unity, this product would be an almost insensible proportion of the total power applied.

From the foregoing, then, it will be perceived that a curved generatrix, so far from being advantageous, is positively a disadvantage; nor is it necessary to depend entirely on induction for this opinion; for it has been fully confirmed by experiments.

About the year 1847 (I give the date from memory) a series of most complete experiments were made by Bourgois, by order of the French government, on a vast number of screws of different proportions and shape, among them the form of screw (*i. e.*, with a curved generatrix) afterwards introduced into the U. S. Navy by Isherwood. These experiments were made with great sagacity of method, and determined most satisfactorily the total uselessness of a curved generatrix. I give the condensed results of the experiments, which will be the first time they have appeared in an English work.

It was perfectly comprehended by Bourgois, that in order to make the influence of a curved generatrix sensible, it would be necessary that the screw have a very considerable slip; its surface, during the experiments, was therefore sufficiently reduced to satisfy this condition.

To fully test this influence, the screw was first tried propelling with the convex face, then tried propelling with the concave face, and lastly tried after the flexure had been taken out of the generatrix; that is, after the generatrix had been made straight.

The results are as follows :—

Time.	Conditions.	Slip of Screw.	
March 4	Strong breeze, river rough, propelling with the convex face of the screw. Mean of six experiments.	per cent. ·50.2	With curved generatrix.
Same day Same time	Propelling with the concave face of the screw. Mean of six experiments.	49.3	
March 19	Calm. Propelling with the convex face of the screw. Mean of six experiments.	44.4	
Same day Same time	Propelling with concave face of the screw. Mean of six experiments.	47.6	
March 20	Calm. Propelling with convex face of the screw. Mean of four experiments.	45.4	
Same day Same time	Propelling with the concave face of the screw. Mean of four experiments.	48.8	
Same day Same time	The same screw having the generatrix made straight. Mean of eight experiments, four being made on each face.	51.2	With straight generatrix.

Allowing for unfavorable errors of observation, dimensions, &c., in experiments of this nature, it will be observed that sensibly the same result was obtained, propelling with either face of the curved generatrix, and the straight generatrix, showing that the employment of a curved generatrix was at least useless, even

with the exaggerated **slip of 50 per** cent.; 30
per cent. being the **maximum in** practice.

After reviewing **some** experiments **carefully
made on** other screws, for the **purpose of** determining **the** effect of a curved generatrix,
Bourgois remarks, which I translate, as follows:—

In the second place, if **we** observe **helicoid
at** surfaces with **curved** generatrices, or what
amounts to the same thing, generated by a
straight line inclined on the axis, we perceive
the liquid thread **does not rest on** the same helicoidal thread.* **As the** periphery of the screw
is approached, the helicoidal **thread** inclines
itself slightly to the centre; there, **on the** contrary, and for the same reason, the liquid threads
tend to remove themselves from the **axis,** but
being endowed with less momentum than the
first, there results **a flowing of** water towards
the centre **of the screw, with so** much the more
abundance **as the curvature of the** generatrix
is greater. This is **the** only notable effect
resulting from **the** employment of a curved
generatrix; **and** there is nothing to **prove** that
effect favorable.

---

* The curvature of the generatrix was not proportioned
to the centrifugal force or **to** the slip.—N.

In the experiments made on screws, B 7 and B 8* show, on the contrary, an increase of slip of 04.7, proving that the employment of curved generatrices directs the water towards the axis in consequence of their obliquity.

The experiments on screws ϕ_3 and ϕ_4 under similar conditions, gave sensibly the same result, either the water was pushed towards the axis or it was deflected out towards the periphery.†

Finally, in passing from screw Ω to Ω_1, the slip increased 07.5 per cent.‡

The experiments of Sabloukoff on a screw turned in air, and having the phenomena made

---

* Which were precisely alike, and formed with the generatrix tangent to an inner cylinder, which is virtually a curved generatrix. B 7 propelled with the obtuse or convex face, and gave a slip of 26.1 per cent. B 8 propelled with an acute or concave face, and gave a slip of 30.8 per cent., or 4.7 per cent. more.—B. F. I.

† Screws ϕ_3 and ϕ_4 were precisely alike, and formed with the generatrix tangent to an inner cylinder. Screw ϕ_3 propelled with the obtuse or convex face, and gave a slip of 32.8 per cent. Screw ϕ_4 propelled with the acute or concave face, and gave a slip of 33 per cent.—B. F. I.

‡ Both screws being precisely alike, except that screw Ω had a straight generatrix, and gave a slip of 34.5 per cent., while screw Ω_1 had a curved generatrix, and gave a slip of 42 per cent.—B. F. I.

.visible by smoke, also **corroborates** the above. It was found that **after a high** rotary speed had **been given to** the screw, the smoke **being** then let **on at its anterior** extremity at any **point** *near its periphery*, was drawn towards the screw, and carried towards the other extremity. When let on **at its** anterior extremity *near the axis*, the smoke **coursed** along parallel to the **axis**, without **any** appearance of having any **circular** movement, **and which was the** same in **the first case; spreading out** from the axis, which **should have been the case had** the rotation of the **screw** been able to **give a** *sensible* centrifugal force to the smoke.

But even supposing (which **we have seen is** far **from** being the **case),** that **the** centrifugal force communicated to **the** particles of water **in contact with the propelling** surface by its rotary movement, **were great enough** to produce so sensible a **result as a vacuum** at the axis of the screw equal to **a diameter one-fourth** the diameter of **the screw, and** supposing the use of a · curved **generatrix to** wholly obviate this, or restore solid water in the place **of the** vacuum; **even** then the employment **of a curved** generatrix would **be** useless as far **as** its reduction of the slip of the screw **is** concerned, and this **fact** also depends on **the carefully** conducted **ex-**

periments of Bourgois. In those experiments there were tried two screws, exactly alike, excepting that the one had a projected area at right angle to the axis of 187.86, while the other had a similar area of 182.59, the reduction being made by cutting out the surface immediately around the axis. The diameter of the screw was 15.752, and the diameter of the cut out part of the last screw was 3.938. The slip of the first was 35.2 per cent., of the last 32.6 per cent.

Similar experiments on two other screws, differing from the above in pitch only, gave with the full screw a slip of 26.9 per cent., with the cut out screw 24.4 per cent. On these experiments Bourgois remarks, which I translate as follows:—

"The difference (between the slip) being little enough to be attributed to irregularities of construction or slight errors in the observations, nothing could be concluded from it, except that a hollowing out, of which the diameter is equal to the fourth part of the exterior diameter of the screw, has no influence on the slip.

"Believing that sufficient has been written *to put this subject in a true light*, there only remains to notice that, when the date of Bourgois' ex-

13

periments are considered so far antecedent to
Isherwood's screw, **it is really** amazing that our
Navy Department **should have** adopted a propeller without **novelty;** but as I apprehend no
one will **ever** use it outside of the navy, the
mischief will **do no** further harm.

In order to prove beyond a shadow of doubt
that the substance **of the** above argument is
based upon a solid **foundation, I beg** to refer to
the highest authority of the land; namely, Mr.
Isherwood's Contribution to *Journal of the
Franklin Institute* for July, 1851, page 42, &c.

A PROPELLER AS CONSTRUCTED FROM MR. ISHER-
WOOD'S DRAWINGS. (PLATE V.)

Having commented **upon** the propeller, as
constructed **in the Bureau of** Steam-Engineering, and represented **by** Plate **IV.,** it is now
proposed **to** describe how propellers have been
constructed **for the navy** by contractors in private establishments.

The specifications of the **construction of** propellers made in the Bureau of Steam-Engineering, although perfectly correct, were not sufficiently clear to enable **the** contractors to follow
the drawings; for which reason, a great many

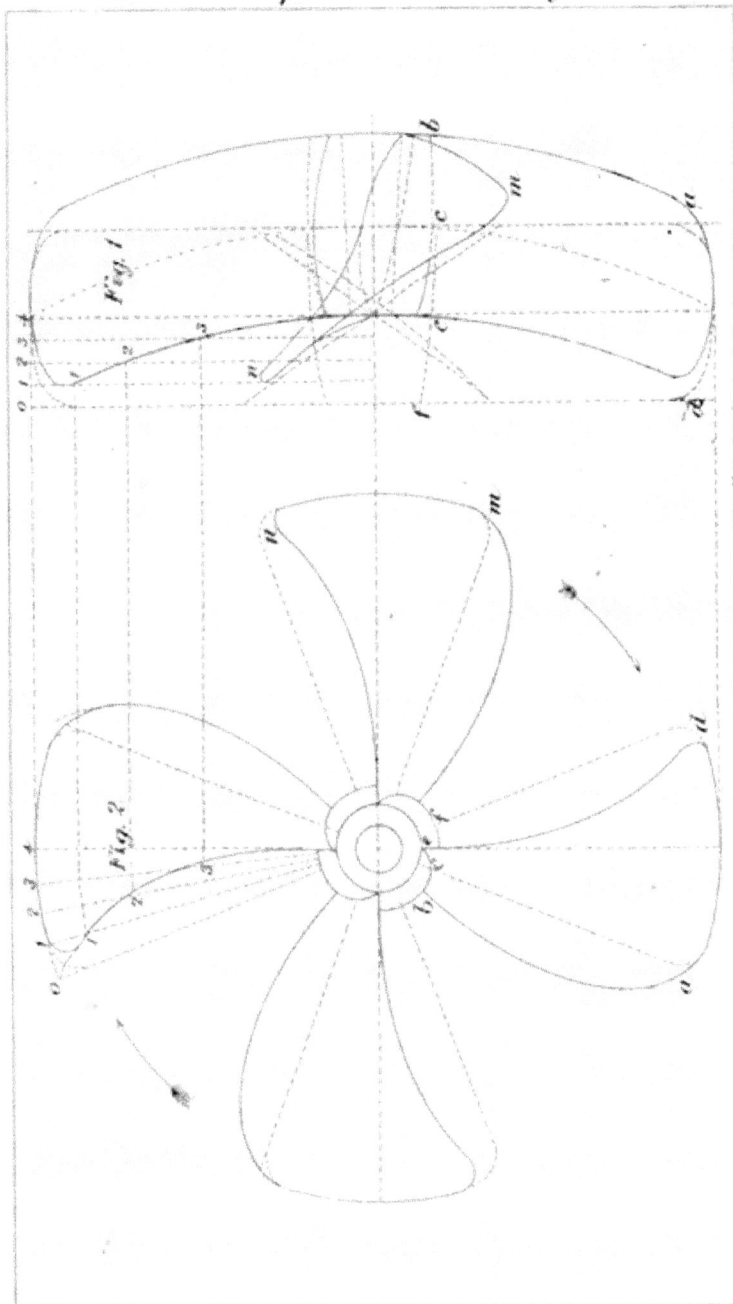

Fig. 1

Fig. 2

propellers have been constructed as represented by Plate V., which appear to have a curved generatrix in both figs. 1 and 2, but in reality the generatrix is a straight line at right angles with the axis, as represented by the dotted lines.* Take off the part *a b c*, and put it on at *d e f*, without changing the direction of the helicoidal surface, the propeller will be the same as represented by Plate II., or a common propeller with straight generatrix. The propelling efficiency of both the propellers on Plates II. and V. will be alike. Viewing the propeller at fig. 2, Plate V., it appears as if the blades would assist the centrifugal force in throwing the water out, which is not the case.

The only advantage of this propeller is, that it will not shake the vessel so much as those represented in Plates II., III., and IV., but its propelling efficiency is some ten per cent. less than that of Plate IV.

A great many propellers of Plate V. were made for naval vessels, and on one occasion the writer remarked in a private establishment that the propeller then moulding was not rightly constructed, or was not according to the drawing furnished by the Navy Department, when

---

* A large brass propeller of this construction is now lying in the Washington Navy Yard, probably condemned. March 1, 1866.

a discussion arose, which resulted in stopping the moulding of the propeller, and the making of a new pattern according to the drawing.

Perhaps some of these errors, the extent and character of which can only be detected and calculated by the application of scientific principles, may serve to explain why it is that the speed of vessels in the navy is so unsatisfactory.

TO CONSTRUCT A CENTRIPETAL PROPELLER OF A UNIFORM PITCH. (PLATE VI.)

Let us now forget all that has been said about the propellers of curved generatrix, and start on an entirely new basis.

The water acted upon by a straight-bladed propeller, is thrown out radially towards the periphery by the action of the centrifugal force; in which case dynamic effect is evidently expended in giving this motion to the water, and as the direction of the motion is at right angles with that of the vessel, the effect expended upon it is thrown away. It is, therefore, now proposed to construct the helicoidal surface of the propeller blades, so as to utilize this lost effect and prevent the water from being thrown out by the centrifugal force.

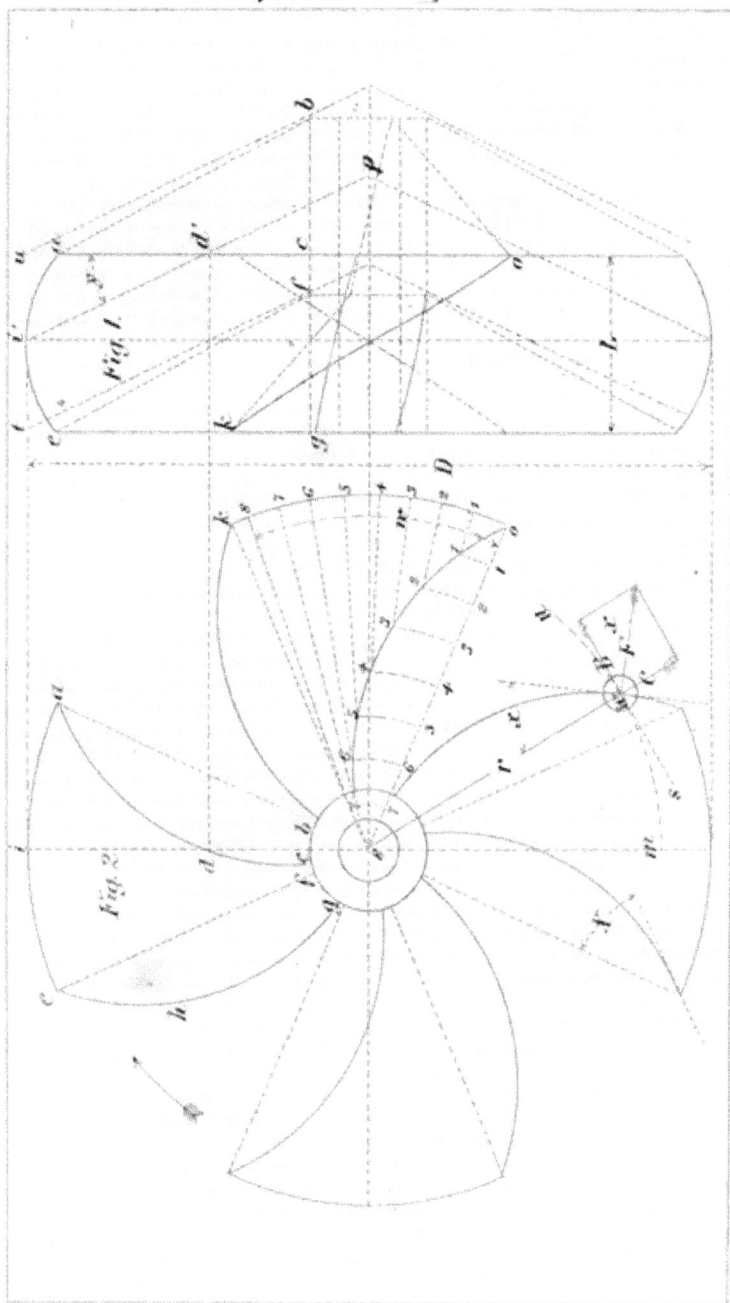

Fig. 1.

Fig. 2.

Let *W*, fig. 2, Plate **VI.**, represent a drop of water acted upon by a force whose magnitude and direction are represented by the arrow *B*, at right angles with the radius *r*. If acted upon by no other force, the drop *W* would move in the direction of *B* towards *s*, but as it is desired to move the water in the direction of the circle *n m*, or more correctly, in the direction of the helix of the screw, it will be necessary to apply a centripetal force, whose magnitude and direction may be represented by the arrow *C*, in the direction of the radius *r*. If the centripetal force *C* be equal to the centrifugal force of the water, then the combined action of the two forces *B* and *C*, would move the drop *W* in the direction of the circle *n m*. The resultant of *B* and *C* may be represented by the magnitude and direction of the arrow *F*, which is the diagonal of the rectangle of *B* and *C*. If the direction of the generatrix of the propeller blade was at right angles with *F*, it would drive the drop *W* in the direction of the circle *n m*, as desired.

The Centrifugal and Centripetal Forces

can be represented by the formula

$$C = \frac{W v^2}{g\, r}. \qquad . \qquad . \qquad 4$$

13*

in which letters denote

W = weight of the drop of water, fig. 2.

v = velocity in feet per second of W.

r = radii in feet of the circle $n\,m$.

g = 32.166, the acceleratrix of gravity.

C = centripetal force, expressed in the same units of weight as W.

n = number of revolutions per minute of the propeller.

$$v = \frac{2\pi r n}{60}. \qquad \cdot \qquad \cdot \qquad \cdot \quad 5$$

Insert formula 5 for v in formula 4, we have

$$C = \frac{W 4\pi^2 r^2 n^2}{60^2 g\, r} = \frac{W 4\pi^2 r n^2}{60^2 g}. \qquad \cdot \qquad \cdot \quad 6$$

Let the propelling force B represent the magnitude and direction of the water W. Then

$$W : C = 1 : tang.\ x.$$

or

$$C = W\ tang.\ x. \qquad \cdot \qquad \cdot \qquad \cdot \quad 7$$

but

$$C = \frac{W 4\pi^2 r n^2}{60^2 g} = W\ tang.\ x.$$

of which

$$tang.\ x \frac{4\pi^2 r n^2}{60^2 g}. \qquad \cdot \qquad \cdot \qquad \cdot \quad 8$$

This formula 8 gives the angle of the generatrix to the radii r.

Let X be the angle of the generatrix at the extremity of the blades, we have

$$tang. X = \frac{2 \pi D n^2}{60^2 g} = \frac{D n}{5870}. \qquad . \qquad . \qquad . \quad 9$$

The centripetal generatrix will be an arithmetic spiral of the angle X at the periphery.

$$tang. X = \frac{\pi w^\circ}{180}$$

when w° is the angles in degrees, in which the spiral is constructed, as shown in fig. 2.

$$tang. X = \frac{\pi w^\circ}{180} = \frac{D n^2}{5870}$$

$$w^\circ = \frac{180 \, D n^2}{5870 \, \pi} = \frac{D n^2}{102.4}. \qquad . \qquad . \quad 10$$

This formula 10, will give the proper angle w° if the slip of the propeller is unity, but the number of revolutions n must be multiplied by the slip S expressed in a fraction of unity, or

$$w^\circ = \frac{D n^2 S^2}{102.4}. \qquad . \qquad . \quad 11$$

From this formula 11, calculate the centripetal angle w° fig. 2, Plate VI. Divide the arc and the radius into any number of (say eight) equal parts and construct the arithmetic spiral as shown by the fig. 2. This spiral will then be centripetal under the condition of formula 11, that is to say, the water will **not** be thrown out by the centrifugal force.

· A curved generatrix at right angles with the axis will form the same helicoidal surface as a straight generatrix inclined to the axis, as before stated.

From the point d, where the dotted line $i\,d$. intersects the generatrix $a\,d\,c$, draw the line $d\,d'$ parallel to the axis of the propeller fig. 1; join $i'\,d'$, continued to p, then $i'\,p$ is the inclined generatrix which will generate the same helicoidal surface as the curved one $a\,d\,c$.

The inclination of the generatrix will be

$$tang.\,y = \frac{w^\circ\,p}{180\,\text{D}}.\qquad \cdot \quad \cdot \quad 12$$

But $w^\circ = \dfrac{\text{D}\,n^2\,S^2}{102.4}$. and $tang.\,y = \dfrac{\text{D}\,n^2\,S^2\,P}{102.4 \times 180\,\text{D}}$

$$tang.\,y = \frac{n^2\,S^2\,P}{18432}.\qquad \cdot \quad \cdot \quad \cdot \quad 13$$

in which P must be expressed in feet.

Draw $u\,b$ and $t\,f$ parallel with $i'\,p$, project a and e, as shown in the figures. Draw from the corners of the blades, fig. 2, the dotted lines to the centre; the dotted lines will represent a propeller with apparently straight blades in fig. 2, and with an inclined generatrix in fig. 1.

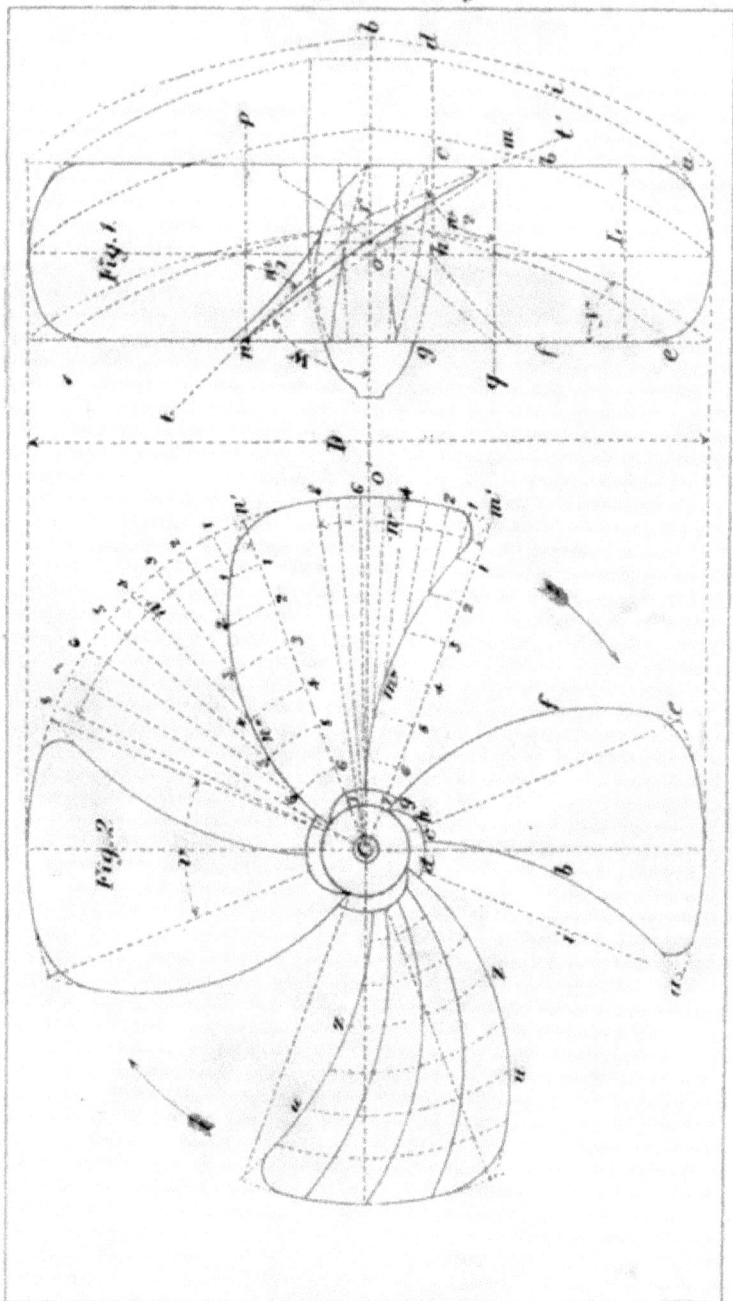

Fig. 1.

Fig. 2.

Both these propellers will produce the same propulsive effect, but that with curved blades indicated by the dark lines in the drawing will not shake the vessel so much as the other shown by the dotted lines.

Captain John Ericsson makes his propellers as shown by the dotted lines, which in reality is a centripetal propeller with a curved generatrix at right angles to the axis.

TO CONSTRUCT A CENTRIPETAL PROPELLER WITH A COMPOUND EXPANDING PITCH. (PLATE VII.)

Having given the diameter D, pitch P, and length L, of the propeller, calculate the angles W, w_1 and w_2 by formula 2. Construct the outer edge of the blade as described for the propeller on Plate III. Calculate the centripetal angle $w°$ from formula 11, for the mean pitch of the propeller. The projecting angle of the blades will be—

$$v = \frac{360\ L}{P}, \qquad \cdot \quad \cdot \quad \cdot \quad 14$$

Make the centripetal angle of the leading generatrix $m'\ m''\ C$,

$$w = w° - \frac{v\ S}{2}.$$

Make the centripetal angle of the delivery generatrix $n'\ n''\ C$,

$$w' = w° + \frac{v\ S}{2}.$$

Construct the two generatrices forming the sides of the blades, which will then constitute a centripetal propeller with a compound expanding pitch.

In making the pattern, or in the moulding of this propeller, it is best to construct several (say five) generatrices, as shown in one blade, which is accomplished by dividing the dotted arcs $u\ u$ and $z\ z$, each into four equal parts, which form the required generatrices.

This is the propeller which the writer would recommend as the best.

Let the helicoidal surface of the propeller be projected backwards at the hub, as shown by the dotted lines, fig. 1, so that the part $e\ f\ g\ h$ be removed to $a\ b\ c\ d\ i$, then the propeller blades would appear straight in fig. 2, though the helicoidal surface and propelling efficiency would be the same in both cases.

The propeller represented by the dotted lines is nearly the same as that on Plate IV., as constructed by Mr. Isherwood. The propeller on Plate VII. is constructed on true scientific principles, which is not the case with the one on Plate IV.

It may be well to explain to the readers who are not familiar with the subject, that tho remarks on Mr. Isherwood's curve-bladed propeller (Plate IV.) commencing at page 133, are his own, as applied to the author's propeller (Plate VII.), and a transposition of names is therefore necessary to make the sense intelligible. See *Journal of the Franklin Institute*, July, 1851.

It indicates what pains the chief took in condemning the curve-bladed propeller, "believing," as he said, "that sufficient has been written to put this subject in a true light." I was very much obliged to the chief for the clearness of his true light, and sincerely hoped its brilliancy would not serve merely to make darkness visible, but the hope has not been realized.

Soon after, Mr. Isherwood constructed the San Jacinto propeller, which turned out a failure, and disgrace to the nation, at her arrival in Constantinople, as the indignant correspondence of Americans from that place abundantly testified.

On the return of this frigate, Mr. Isherwood's propeller was condemned, and a better one from a private establishment was substituted.

When Mr. Isherwood became more enlightened on the subject, he found that the curve-bladed propeller **was all** right, and quietly adopted it in the navy.

His empiricism thus triumphed not only over **me**, but over the navy **and** the nation, and the country has been most severely injured by it too. True scientific principles applied by a civilian, **have** been attacked and vanquished by quackery from the navy.

Neglecting all **science** and theory, experience **alone** has led to the adoption of curve-bladed propellers. I **have made** experiments with **a** great many different kinds of screws, in which the powers expended and delivered were **correctly** measured by a delicate dynamometer, and which indicated a decided advantage on the side of the curve-blades. In 1846 I proposed a curve-bladed propeller to Captain Carlsund, at Motala, Sweden, who rejected it **for the** reason that it would not back so well as **the** straight-bladed **one**. Captain Carlsund, however, has since adopted the curve-screw exclusively.

The French experiments quoted by Mr. Isherwood, I am inclined to believe, are not reliable, inasmuch as they are at variance with subsequent experience, and declare a preference for convex over concave surfaces in propelling.

There **is a** propeller introduced in this

country, called the "buffalo wheel," in which
the curve is turned the wrong way, or in other
words, propels with a convex surface. I have
been on board of several steamers with this
screw, and the engineers have invariably told
me that it works better in backing than ahead,
which confirms the principles herein given,
and conflicts with the results of the French
experiments.

Some years ago there was a steamer built in
Chicago, Ill., with two propellers. It was decided to put a common straight-bladed propeller on one side, and a curve-bladed on the
other. After she had been running for some
time, I received an order for another curve-
bladed propeller to take the place of the
straight-bladed one, in consequence of the su-
periority of the former.

In a great many steamers in the navy there
is not room enough between the stern and
rudder posts to admit the drawn propeller
represented on Plate IV., but there is room for
the dotted one, which is equally efficient in
propelling, and, still better, the drawn propeller represented on Plate VII. But in so
doing, Mr. Isherwood's empiricism might be
exposed in regard to curve bladed propellers,
although it would not be necessary to infringe

14

upon the centripetal propeller and make the generatrix a true arithmetic spiral which it ought to be; and in order to avoid the risk of science he can easily make a quack spiral, which would still make a good propeller.

Mr. Isherwood hastily and erroneously committed himself in the propeller question, precisely as in the anti-expansion question, so that now he cannot act according to his own convictions, but constrains the navy to suffer the consequences.

In all this, Mr. Isherwood is not so much to be blamed as the custom by which the profession has so long been guided, and from which it has suffered such serious evils. It is lamentable for the country to ruin such extraordinary talent as that with which Mr. Isherwood is naturally gifted.

Under the present organization, experience has demonstrated that it matters little who is the engineer-in-chief, for he is so overruled by politics that he cannot, even with all his goodwill, act altogether advantageously for his office.

The Bureau of Steam-Engineering has been required by Congress to adopt machinery so perfectly absurd, that not a shadow of success for it could have been anticipated. After im-

mense sums of money have been expended,
and the failure proclaimed, then the engineer-
in-chief, as well as the navy department, have
been attacked and shamefully abused for what
they have not been at fault; as has also been
the case in the notorious trials of the steamers
Algonquin and Winooski.

The people know only what has been published in the newspapers, where it is impossible
to separate the chaff from the grain, and the
true state of the case has not yet been revealed.
In the course of this protracted controversy,
however, our engineering standing has been
impaired, and great loss of money has ensued;
not from any want of talent in the Bureau of
Steam-Engineering, but simply because it does
not control the confidence, and command the
respect and dignity due to its important office.

Take the office of the Coast Survey as an example. The chief there not only understands
his business, but he is master of his situation.
Congress will not impose upon him the adoption of some gimcrack instrument in his surveying. We never hear complaints of the
maps and reports of that department, which
command respect throughout the world, and
are second to none of their kind. On the one

side there is dignity and learning, and on the other, pedantry and dogmatism.

The engineer-in-chief of the U. S. navy ought to be endowed with the highest rank of that department. He ought to be brought up from some properly established technological academy, through all the different branches of naval engineering, including experience in the workshops and yards, and even in the coast-survey, lighthouse board and observatory departments, all of which are proper appurtenants of the navy.

When an engineer has thus reached the important and responsible station of engineer-in-chief, he would be able to command all the respect and confidence due to so distinguished an office, and in intellectual rank would be equivalent to a Grand Admiral. He should be the engineer-in-chief not only for "engine driving," but for the yards and docks, constructions and works of every kind. He would himself be far above the drudgery and detail of mere construction, but would intrust that to the commodore-engineer in each yard where the work is to be executed, thus giving a chance of development and display to whatever talent the corps of engineers might possess, and create an emulation which would elevate the navy to a condition of the highest perfection.

The Grand Admiral Engineer would know how to select and surround himself with the highest ability, and how to detail appropriate persons to their respective stations. He would certainly build no light draft monitors which would not float. He would make no anti-expansion experiments and researches in steam-engineering without consulting the physical laws involved in the operation. He would not build any *gimcrack-hair-cut-off-antifriction-double-double-crank-machinery.* There would be no pamphlets or newspapers abusing the engineer-in-chief and the navy department in general. The seed of technological education would be realized in a valuable harvest, and give no occasion for attacks upon high officials "who have not the *savoir faire* to chop up an opponent without hurting his feelings."

As it now stands, the chief takes upon him-self to restrict all construction to his individual notions (except so far as he is himself controlled by the politicians), and from a censurable ambi-tion, is afraid to endow with discretion any of his subordinates whom he suspects of talent which may surpass his own.

Constructions ought never to be made in the

navy department, for engineers, not being immediately connected with the workshops, cannot keep pace with the practical progress which is going on there.

When Mr. Isherwood entered the U.S. navy, his natural talent for engineering made him at once his own master. There was none above him whose distinction he feared, none having sufficient technological education to control or analyze his reasonings, and restrain the amazing impetuosity which characterized all his movements.

Now, in a properly educated and well organized corps of engineers, Mr. Isherwood would have been subjected to such a wholesome supervision that his great talent would have been usefully developed and utilized, and most of his well-meant errors would not have occurred to the detriment of his own rising reputation, and to the damage of that *esprit de corps* which we all are so desirous to encourage.

The technological academy, as above hinted, should also embrace ordnance, coast survey, and lighthouse engineering, all of which naturally belong to the navy, and ought to be superintended by naval engineers. Lighthouse engineering presents a wide field, but is yet very little studied. The construction of lighthouses and lightships with their appurtenances, as lamps, lenses, reflectors, electric lights, and the different kinds of machinery connected therewith, requires great mechanical skill, and ought to be the work of naval engineers.

The navy yards, which are now under the charge of line-officers, ought to be intrusted to engineers of the same rank. In private life, we never find a shipyard or machine shop in charge of a sea-captain. A carpenter cannot superintend the work of a blacksmith.

The magnificent combination of a properly organized corps of naval engineers would at once elevate the country as well as economize its means, and utilize its resources.

The money lost or squandered during the rebellion for want of such a corps may be estimated at a *hundred million of dollars*, the interest of which, through all future time, would be more than sufficient to build and support a technological academy of the highest order, and pay the salary of the whole corps of engineers.

Who can predict the coming destiny of the country? Who can tell how soon we may have another protracted war? Are we prepared to meet it without extravagant sacrifices?

There is no doubt of our capability to defeat any enemy that would dare to meet us, but the

question is, not to waste our means and ammu-
nition at random, or to give him the satisfaction
of knowing that we have overstrained ourselves
in the conflict, but to convince him on the contrary that *our power* is a manifestation of skill,
and is not measured by numbers of guns and
dollars.

Great discoveries are frequently made of the
highest national importance, and we are unable
to grasp hold of them for want of a competent
bureau of technical knowledge.

Take, for example, the **Bessemer** process of
refining iron, which, although announced in
England some ten years ago, and although
every nation in Europe took hold of it at once,
there is yet but one establishment of the kind,
and that only recently erected, in the United
States. Had we been less dilatory, and secured
the immense resources its introduction would
have given us, it would have saved us many
thousands of lives, and many millions of dollars,
during our late naval and military operations.

CATALOGUE

OF

PRACTICAL AND SCIENTIFIC BOOKS,

PUBLISHED BY

HENRY CAREY BAIRD,

Industrial Publisher,

NO. 406 WALNUT STREET,

PHILADELPHIA.

☞ Any of the Books comprised in this Catalogue will be sent by mail, free of postage, at the publication price.

☞ A Descriptive Catalogue, 96 pages, 8vo., will be sent, free of postage, to any one who will furnish the publisher with his address.

ARLOT.—A Complete Guide for Coach Painters.
Translated from the French of M. ARLOT, Coach Painter; for eleven years Foreman of Painting to M. Eherler, Coach Maker, Paris. By A. A. FESQUET, Chemist and Engineer. To which is added an Appendix, containing Information respecting the Materials and the Practice of Coach and Car Painting and Varnishing in the United States and Great Britain. 12mo. $1.25

ARMENGAUD, AMOROUX, and JOHNSON.—The Practical Draughtsman's Book of Industrial Design, and Machinist's and Engineer's Drawing Companion:
Forming a Complete Course of Mechanical Engineering and Architectural Drawing. From the French of M. Armengaud the elder, Prof. of Design in the Conservatoire of Arts and Industry, Paris, and MM. Armengaud the younger, and Amoroux, Civil Engineers. Rewritten and arranged with additional matter and plates, selections from and examples of the most useful and generally employed mechanism of the day. By WILLIAM JOHNSON, Assoc. Inst. C. E., Editor of "The Practical Mechanic's Journal." Illustrated by 50 folio steel plates, and 50 wood-cuts. A new edition, 4to. $10.00

1

ARROWSMITH.—Paper-Hanger's Companion:

A Treatise in which the Practical Operations of the Trade are Systematically laid down: with Copious Directions Preparatory to Papering; Preventives against the Effect of Damp on Walls; the Various Cements and Pastes Adapted to the Several Purposes of the Trade; Observations and Directions for the Panelling and Ornamenting of Rooms, etc. By JAMES ARROWSMITH, **Author** of "Analysis of Drapery," etc. 12mo., cloth. $1.25

ASHTON.—The Theory and Practice of the Art of Designing Fancy Cotton and Woollen Cloths from Sample:

Giving full Instructions for Reducing Drafts, as well as the Methods of Spooling and Making out Harness for Cross Drafts, and Finding any Required Reed, with Calculations and Tables of Yarn. By FREDERICK T. ASHTON, Designer, West Pittsfield, Mass. With 52 Illustrations. **One** volume, 4to. $10.00

BAIRD.—Letters on the Crisis, the Currency and the Credit System.

By HENRY CAREY BAIRD. Pamphlet.   05

BAIRD.—Protection of Home Labor and Home Productions necessary to the Prosperity of the American Farmer.

By HENRY CAREY BAIRD. 8vo., paper.   10

BAIRD.—Some of the Fallacies of British Free-Trade Revenue Reform.

Two Letters to Arthur Latham Perry, Professor of History and Political Economy in Williams College. By HENRY CAREY BAIRD. Pamphlet.   05

BAIRD.—The Rights of American Producers, and the Wrongs of British Free-Trade Revenue Reform.

By HENRY CAREY BAIRD. Pamphlet.   05

BAIRD.—Standard Wages Computing Tables:

An Improvement in all former Methods of Computation, so arranged that wages for days, hours, or fractions of hours, at a specified rate per day or hour, may be ascertained at a glance. By T. SPANGLER BAIRD. Oblong folio. $5.00

BAIRD.—The American Cotton Spinner, and Manager's and Carder's Guide:

A Practical Treatise on Cotton Spinning; giving the Dimensions and Speed of Machinery, Draught and Twist Calculations, etc.; with notices of recent Improvements: together with Rules and Examples for making changes in the sizes and numbers of Roving and Yarn. Compiled from the papers of the late ROBERT H. **BAIRD.** 12mo. $1.50

BAKER.—Long-Span Railway Bridges :

Comprising Investigations of the Comparative Theoretical and Practical Advantages of the various Adopted or Proposed Type Systems of Construction ; with numerous Formulæ and Tables. By B. BAKER. 12mo. $2.00

BAUERMAN.—A Treatise on the Metallurgy of Iron :

Containing Outlines of the History of Iron Manufacture, Methods of Assay, and Analysis of Iron Ores, Processes of Manufacture of Iron and Steel, etc., etc. By H. BAUERMAN, F. G. S., Associate of the Royal School of Mines. First American Edition, Revised and Enlarged. With an Appendix on the Martin Process for Making Steel, from the Report of ABRAM S. HEWITT, U. S. Commissioner to the Universal Exposition at Paris, 1867. Illustrated. 12mo. . $2.00

BEANS.—A Treatise on Railway Curves and the Location of Railways.

By E. W. BEANS, C. E. Illustrated. 12mo. Tucks. . . $1.50

BELL.—Carpentry Made Easy :

Or, The Science and Art of Framing on a New and Improved System. With Specific Instructions for Building Balloon Frames, Barn Frames, Mill Frames, Warehouses, Church Spires, etc. Comprising also a System of Bridge Building, with Bills, Estimates of Cost, and valuable Tables. Illustrated by 38 plates, comprising nearly 200 figures. By WILLIAM E. BELL, Architect and Practical Builder. 8vo. . $5.00

BELL.—Chemical Phenomena of Iron Smelting :

An Experimental and Practical Examination of the Circumstances which determine the Capacity of the Blast Furnace, the Temperature of the Air, and the proper Condition of the Materials to be operated upon. By I. LOWTHIAN BELL. Illustrated. 8vo. . . $6.00

BEMROSE.—Manual of Wood Carving :

With Practical Illustrations for Learners of the Art, and Original and Selected Designs. By WILLIAM BEMROSE, Jr. With an Introduction by LLEWELLYN JEWITT, F. S. A., etc. With 128 Illustrations. 4to., cloth. $3.00

BICKNELL.—Village Builder, and Supplement :

Elevations and Plans for Cottages, Villas, Suburban Residences, Farm Houses, Stables and Carriage Houses, Store Fronts, School Houses, Churches, Court Houses, and a model Jail ; also, Exterior and Interior details for Public and Private Buildings, with approved Forms of Contracts and Specifications, including Prices of Building Materials and Labor at Boston, Mass., and St. Louis, Mo. Containing 75 plates drawn to scale ; showing the style and cost of building in different sections of the country, being an original work comprising the designs of twenty leading architects, representing the New England, Middle, Western, and Southwestern States. 4to. . $12.00

BLENKARN.—Practical Specifications of Works executed in Architecture, Civil and Mechanical Engineering, and in Road Making and Sewering:

To which are added a series of practically useful Agreements and Reports. By JOHN BLENKARN. Illustrated by 15 large folding plates. 8vo. $9.00

BLINN.—A Practical Workshop Companion for Tin, Sheet-Iron, and Copperplate Workers:

Containing Rules for describing various kinds of Patterns used by Tin, Sheet-Iron, and Copper-plate Workers; Practical Geometry; Mensuration of Surfaces and Solids; Tables of the Weights of Metals, Lead Pipe, etc.; Tables of Areas and Circumferences of Circles; Japan, Varnishes, Lackers, Cements, Compositions, etc., etc. By LEROY J. BLINN, Master Mechanic. With over 100 Illustrations. 12mo. $2.50

BOOTH.—Marble Worker's Manual:

Containing Practical Information respecting Marbles in general, their Cutting, Working, and Polishing; Veneering of Marble; Mosaics; Composition and Use of Artificial Marble, Stuccos, Cements, Receipts, Secrets, etc., etc. Translated from the French by M. L. BOOTH. With an Appendix concerning American Marbles. 12mo., cloth. $1.50

BOOTH AND MORFIT.—The Encyclopedia of Chemistry, Practical and Theoretical:

Embracing its application to the Arts, Metallurgy, Mineralogy, Geology, Medicine, and Pharmacy. By JAMES C. BOOTH, Melter and Refiner in the United States Mint, Professor of Applied Chemistry in the Franklin Institute, etc., assisted by CAMPBELL MORFIT, author of "Chemical Manipulations," etc. Seventh edition. Royal 8vo., 978 pages, with numerous wood-cuts and other illustrations.   . $5.00

BOX.—A Practical Treatise on Heat:

As applied to the Useful Arts; for the Use of Engineers, Architects, etc. By THOMAS BOX, author of "Practical Hydraulics." Illustrated by 14 plates containing 114 figures. 12mo. $4.25

BOX.—Practical Hydraulics:

A Series of Rules and Tables for the use of Engineers, etc. By THOMAS BOX. 12mo. $2.50

BROWN.—Five Hundred and Seven Mechanical Movements:

Embracing all those which are most important in Dynamics, Hydraulics, Hydrostatics, Pneumatics, Steam Engines, Mill and other Gearing, Presses, Horology, and Miscellaneous Machinery; and including many movements never before published, and several of which have only recently come into use. By HENRY T. BROWN, Editor of the "American Artisan." In one volume, 12mo.   .   .   . $1.00

BUCKMASTER.—The Elements of Mechanical Physics:

By J. C. BUCKMASTER, late Student in the Government School of Mines; Certified Teacher of Science by the Department of Science and Art; Examiner in Chemistry and Physics in the Royal College of Preceptors, and late Lecturer in Chemistry and Physics of the Royal Polytechnic Institute. Illustrated with numerous engravings. In one volume, 12mo. $1.50

BULLOCK.—The American Cottage Builder:

A Series of Designs, Plans, and Specifications, from $200 to $20,000, for Homes for the People; together with Warming, Ventilation, Drainage, Painting, and Landscape Gardening. By JOHN BULLOCK, Architect, Civil Engineer, Mechanician, and Editor of "The Rudiments of Architecture and Building," etc., etc. Illustrated by 75 engravings. In one volume, 8vo. $3.50

BULLOCK.—The Rudiments of Architecture and Building:

For the use of Architects, Builders, Draughtsmen, Machinists, Engineers, and Mechanics. Edited by JOHN BULLOCK, author of "The American Cottage Builder." Illustrated by 250 engravings. In one volume, 8vo. $3.50

BURGH.—Practical Illustrations of Land and Marine Engines:

Showing in detail the Modern Improvements of High and Low Pressure, Surface Condensation, and Super-heating, together with Land and Marine Boilers. By N. P. BURGH, Engineer. Illustrated by 20 plates, double elephant folio, with text. . . . $21.00

BURGH.—Practical Rules for the Proportions of Modern Engines and Boilers for Land and Marine Purposes.

By N. P. BURGH, Engineer. 12mo. $1.50

BURGH.—The Slide-Valve Practically Considered.

By N. P. BURGH, Engineer. Completely illustrated. 12mo. $2.00

BYLES.—Sophisms of Free Trade and Popular Political Economy Examined.

By a BARRISTER (Sir JOHN BARNARD BYLES, Judge of Common Pleas). First American from the Ninth English Edition, as published by the Manchester Reciprocity Association. In one volume, 12mo. Paper, 75 cts. Cloth $1.25

BYRN.—The Complete Practical Brewer:

Or Plain, Accurate, and Thorough Instructions in the Art of Brewing Beer, Ale, Porter, including the Process of making Bavarian Beer, all the Small Beers, such as Root-beer, Ginger-pop, Sarsaparilla-beer, Mead, Spruce Beer, etc., etc. Adapted to the use of Public Brewers and Private Families. By M. LA FAYETTE BYRN, M. D. With illustrations. 12mo. $1.25

BYRN.—The Complete Practical Distiller:

Comprising the most perfect and exact Theoretical and Practical Description of the Art of Distillation and Rectification; including all of the most recent improvements in distilling apparatus; instructions for preparing spirits from the numerous vegetables, fruits, etc.; directions for the distillation and preparation of all kinds of brandies and other spirits, spirituous and other compounds, etc., etc. By M. LA FAYETTE BYRN, M. D. Eighth Edition. To which are added, Practical Directions for Distilling, from the French of Th. Fling, Brewer and Distiller. **12mo.** $1.50

BYRNE.—Handbook for the Artisan, Mechanic, and Engineer:

Comprising the Grinding and Sharpening of Cutting Tools, Abrasive Processes, Lapidary Work, Gem and Glass Engraving, Varnishing and Lackering, Apparatus, Materials and Processes for Grinding and Polishing, etc. By OLIVER BYRNE. Illustrated by 185 wood engravings. In one volume, 8vo. $5.00

BYRNE.—Pocket Book for Railroad and Civil Engineers:

Containing New, **Exact,** and Concise Methods for Laying out Railroad Curves, Switches, Frog Angles, and Crossings; the Staking out of work; Levelling; the Calculation of Cuttings; Embankments; Earth-work, **etc.** By OLIVER BYRNE. 18mo., full bound, pocket-book form $1.75

BYRNE.—The Practical Model Calculator:

For the Engineer, Mechanic, Manufacturer of Engine Work, Naval Architect, Miner, and Millwright. By OLIVER BYRNE. 1 volume, 8vo., nearly 600 pages $4.50

BYRNE.—The Practical Metal-Worker's Assistant:

Comprising Metallurgic Chemistry; the Arts of Working all Metals and Alloys; Forging of Iron and Steel; Hardening and Tempering; Melting and Mixing; Casting and Founding; Works in Sheet Metal; The Processes Dependent on the Ductility of the Metals; Soldering; and the most Improved Processes and Tools employed by Metal-Workers. With the Application of the Art of Electro-Metallurgy to Manufacturing Processes; collected from Original Sources, and from the Works of Holtzapffel, Bergeron, Leupold, Plumier, Napier, Scoffern, Clay, Fairbairn, and others. By OLIVER BYRNE. A new, revised, and improved edition, to which is added An Appendix, containing THE MANUFACTURE OF RUSSIAN SHEET-IRON. By JOHN PERCY, M. D., F.R.S. THE MANUFACTURE OF MALLEABLE IRON CASTINGS, and IMPROVEMENTS IN BESSEMER STEEL. By A. A. FESQUET, Chemist and Engineer. With over 600 Engravings, illustrating every Branch of the Subject. 8vo. $7.00

Cabinet Maker's Album of Furniture:

Comprising a Collection of Designs for Furniture. Illustrated by 48 Large and Beautifully Engraved Plates. In one vol., oblong $5.00

CALLINGHAM.—Sign Writing and Glass Embossing:

A Complete Practical Illustrated Manual of the Art. By JAMES CALLINGHAM. In one volume, 12mo. $1.50

CAMPIN.—A Practical Treatise on Mechanical Engineering:

Comprising Metallurgy, Moulding, Casting, Forging, Tools, Workshop Machinery, Mechanical Manipulation, Manufacture of Steamengines, etc., etc. With an Appendix on the Analysis of Iron and Iron Ores. By FRANCIS CAMPIN, C. E. To which are added, Observations on the Construction of Steam Boilers, and Remarks upon Furnaces used for Smoke Prevention; with a Chapter on Explosions. By R. Armstrong, C. E., and John Bourne. Rules for Calculating the Change Wheels for Screws on a Turning Lathe, and for a Wheelcutting Machine. By J. LA NICCA. Management of Steel, Including Forging, Hardening, Tempering, Annealing, Shrinking, and Expansion. And the Case-hardening of Iron. By G. EDE. 8vo. Illustrated with 29 plates and 100 wood engravings . . . $6.00

CAMPIN.—The Practice of Hand-Turning ·in Wood, Ivory, Shell, etc.:

With Instructions for Turning such works in Metal as may be required in the Practice of Turning Wood, Ivory, etc. Also, an Appendix on Ornamental Turning. By FRANCIS CAMPIN; with Numerous Illustrations. 12mo., cloth $3.00

CAREY.—The Works of Henry C. Carey:

FINANCIAL CRISES, their Causes and Effects. 8vo. paper . 25

HARMONY OF INTERESTS: Agricultural, Manufacturing, and Commercial. 8vo., cloth $1.50

MANUAL OF SOCIAL SCIENCE. Condensed from Carey's "Principles of Social Science." By KATE McKEAN. 1 vol. 12mo. $2.25

MISCELLANEOUS WORKS: comprising "Harmony of Interests," "Money," "Letters to the President," "Financial Crises," "The Way to Outdo England Without Fighting Her," "Resources of the Union," "The Public Debt," "Contraction or Expansion?" "Review of the Decade 1857-'67," "Reconstruction," etc., etc. Two vols., 8vo., cloth $10.00

PAST, PRESENT, AND FUTURE. 8vo. $2.50

PRINCIPLES OF SOCIAL SCIENCE. 3 vols., 8vo., cloth $10.00

THE SLAVE-TRADE, DOMESTIC AND FOREIGN; Why it Exists, and How it may be Extinguished (1853). 8vo., cloth . $2.00

LETTERS ON INTERNATIONAL COPYRIGHT (1867) . 50

THE UNITY OF LAW: As Exhibited in the Relations of Physical, Social, Mental, and Moral Science (1872). In one volume, 8vo., pp. xxiii., 433. Cloth $3.50

CHAPMAN.—A Treatise on Ropemaking:

As Practised in private and public Rope yards, with a Description of the Manufacture, Rules, Tables of Weights, etc., adapted to the Trades, Shipping, Mining, Railways, Builders, etc. By ROBERT CHAPMAN. 24mo. $1.50

COLBURN.—The Locomotive Engine :

Including a Description of its Structure, Rules for Estimating its Capabilities, and Practical Observations on its Construction and Management. By ZERAH COLBURN. Illustrated. A new edition. 12mo. $1.25

CRAIK. — The Practical American Millwright and Miller.

By DAVID CRAIK, Millwright. Illustrated by numerous wood engravings, and two folding plates. 8vo. $5.00

DE GRAFF.—The Geometrical Stair Builders' Guide :

Being a Plain Practical System of Hand-Railing, embracing all its necessary Details, and Geometrically Illustrated by 22 Steel Engravings; together with the use of the most approved principles of Practical Geometry. By SIMON DE GRAFF, Architect. 4to. . $5.00

DE KONINCK.—DIETZ.—A Practical Manual of Chemical Analysis and Assaying :

As applied to the Manufacture of Iron from its Ores, and to Cast Iron, Wrought Iron, and Steel, as found in Commerce. By L. L. DE KONINCK, Dr. Sc., and E. DIETZ, Engineer. Edited with Notes, by ROBERT MALLET, F.R.S., F.S.G., M.I.C.E., etc. American Edition, Edited with Notes and an Appendix on Iron Ores, by A. A. FESQUET, Chemist and Engineer. One volume, 12mo. $2.50

DUNCAN.—Practical Surveyor's Guide :

Containing the necessary information to make any person, of common capacity, a finished land surveyor without the aid of a teacher. By ANDREW DUNCAN. Illustrated. 12mo., cloth. . . . $1.25

DUPLAIS.—A Treatise on the Manufacture and Distillation of Alcoholic Liquors :

Comprising Accurate and Complete Details in Regard to Alcohol from Wine, Molasses, Beets, Grain, Rice, Potatoes, Sorghum, Asphodel, Fruits, etc.; with the Distillation and Rectification of Brandy, Whiskey, Rum, Gin, Swiss Absinthe, etc., the Preparation of Aromatic Waters, Volatile Oils or Essences, Sugars, Syrups, Aromatic Tinctures, Liqueurs, Cordial Wines, Effervescing Wines, etc., the Aging of Brandy and the Improvement of Spirits, with Copious Directions and Tables for Testing and Reducing Spirituous Liquors, etc., etc. Translated and Edited from the French of MM. DUPLAIS, Ainé et Jeune. By M. McKENNIE, M.D. To which are added the United States Internal Revenue Regulations for the Assessment and Collection of Taxes on Distilled Spirits. Illustrated by fourteen folding plates and several wood engravings. **743 pp., 8vo.** $10.00

DUSSAUCE.—A General Treatise on the Manufacture of Every Description of Soap :

Comprising the Chemistry of the Art, with Remarks on Alkalies, Saponifiable Fatty Bodies, the apparatus necessary in a Soap Factory, Practical Instructions in the manufacture of the various kinds of Soap, the assay of Soaps, etc., etc. Edited from Notes of Larmé, Fontenelle, Malapayre, Dufour, and others, with large and important additions by Prof. H. DUSSAUCE, Chemist. Illustrated. In one vol., 8vo. . $10.00

DUSSAUCE.—A General Treatise on the Manufacture of Vinegar:

Theoretical and Practical. Comprising the various Methods, by the Slow and the Quick Processes, with Alcohol, Wine, Grain, Malt, Cider, Molasses, and Beets; as well as the Fabrication of Wood Vinegar, etc., etc. By Prof. H. DUSSAUCE. In one volume, 8vo. . . $5.00

DUSSAUCE.—A New and Complete Treatise on the Arts of Tanning, Currying, and Leather Dressing:

Comprising all the Discoveries and Improvements made in France, Great Britain, and the United States. Edited from Notes and Documents of Messrs. Sallerou, Grouvelle, Duval, Dessables, Labarraque, Payen, René, De Fontenelle, Malapeyre, etc., etc. By Prof. H. DUSSAUCE, Chemist. Illustrated by 212 wood engravings. 8vo. $25.00

DUSSAUCE.—A Practical Guide for the Perfumer:

Being a New Treatise on Perfumery, the most favorable to the Beauty without being injurious to the Health, comprising a Description of the substances used in Perfumery, the Formulæ of more than 1000 Preparations, such as Cosmetics, Perfumed Oils, Tooth Powders, Waters, Extracts, Tinctures, Infusions, Spirits, Vinaigres, Essential Oils, Pastels, Creams, Soaps, and many new Hygienic Products not hitherto described. Edited from Notes and Documents of Messrs. Debay, Lunel, etc. With additions by Prof. H. DUSSAUCE, Chemist. 12mo. $3.00

DUSSAUCE.—Practical Treatise on the Fabrication of Matches, Gun Cotton, and Fulminating Powders.

By Prof. H. DUSSAUCE. 12mo. $3.00

Dyer and Color-maker's Companion:

Containing upwards of 200 Receipts for making Colors, on the most approved principles, for all the various styles and fabrics now in existence; with the Scouring Process, and plain Directions for Preparing, Washing-off, and Finishing the Goods. In one vol., 12mo. . $1.25

EASTON.—A Practical Treatise on Street or Horse-power Railways.

By ALEXANDER EASTON, C. E. Illustrated by 23 plates. 8vo., cloth. $2.00

ELDER.—Questions of the Day:

Economic and Social. By Dr. WILLIAM ELDER. 8vo. . $3.00

FAIRBAIRN.—The Principles of Mechanism and Machinery of Transmission:

Comprising the Principles of Mechanism, Wheels, and Pulleys, Strength and Proportions of Shafts, Coupling of Shafts, and Engaging and Disengaging Gear. By Sir WILLIAM FAIRBAIRN, C.E., LL.D., F.R.S., F.G.S. Beautifully illustrated by over 150 wood-cuts. In one volume, 12mo. $2.50

FORSYTH.—Book of Designs for Headstones, Mural, and other Monuments:

Containing 78 Designs. By JAMES FORSYTH. With an Introduction by CHARLES BOUTELL, M. A. 4to., cloth. $5.00

GIBSON.—The American Dyer:

A Practical Treatise on the Coloring of Wool, Cotton, Yarn and Cloth, in three parts. Part First gives a descriptive account of the Dye Stuffs; if of vegetable origin, where produced, how cultivated, and how prepared for use; if chemical, their composition, specific gravities, and general adaptability, how adulterated, and how to detect the adulterations, etc. Part Second is devoted to the Coloring of Wool, giving recipes for one hundred and twenty-nine different colors or shades, and is supplied with sixty colored samples of Wool. Part Third is devoted to the Coloring of Raw Cotton or Cotton Waste, for mixing with Wool Colors in the Manufacture of all kinds of Fabrics, gives recipes for thirty-eight different colors or shades, and is supplied with twenty-four colored samples of Cotton Waste. Also, recipes for Coloring Beavers, Doeskins, and Flannels, with remarks upon Anilines, giving recipes for fifteen different colors or shades, and nine samples of Aniline Colors that will stand both the Fulling and Scouring process. Also, recipes for Aniline Colors on Cotton Thread, and recipes for Common Colors on Cotton Yarns. Embracing in all over two hundred recipes for Colors and Shades, and ninety-four samples of Colored Wool and Cotton Waste, etc. By RICHARD H. GIBSON, Practical Dyer and Chemist. In one volume, 8vo. . . $12.50

GILBART.—History and Principles of Banking:

A Practical Treatise. By JAMES W. GILBART, late Manager of the London and Westminster Bank. With additions. In one volume, 8vo., 600 pages, sheep $5.00

Gothic Album for Cabinet Makers:

Comprising a Collection of Designs for Gothic Furniture. Illustrated by 23 large and beautifully engraved plates. Oblong . . $3.00

GRANT.—Beet-root Sugar and Cultivation of the Beet.

By E. B. GRANT. 12mo. $1.25

GREGORY.—Mathematics for Practical Men:

Adapted to the Pursuits of Surveyors, Architects, Mechanics, and Civil Engineers. By OLINTHUS GREGORY. 8vo., plates, cloth $3.00

GRISWOLD.—Railroad Engineer's Pocket Companion for the Field:

Comprising Rules for Calculating Deflection Distances and Angles, Tangential Distances and Angles, and all Necessary Tables for Engineers; also the art of Levelling from Preliminary Survey to the Construction of Railroads, intended Expressly for the Young Engineer, together with Numerous Valuable Rules and Examples. By W. GRISWOLD. 12mo., tucks $1.75

GRUNER.—Studies of Blast Furnace Phenomena.

By M. L. GRUNER, President of the General Council of Mines of France, and lately Professor of Metallurgy at the Ecole des Mines. Translated, with the Author's sanction, with an Appendix, by L. D. B. Gordon, F. R. S. E., F. G. S. Illustrated. 8vo. . . . $2.50

GUETTIER.—Metallic Alloys:

Being a Practical Guide to their Chemical and Physical Properties, their Preparation, Composition, and Uses. Translated from the French of A. GUETTIER, Engineer and Director of Foundries, author of "La Fouderie en France," etc., etc. By A. A. FESQUET, Chemist and Engineer. In one volume, 12mo. $3.00

HARRIS.—Gas Superintendent's Pocket Companion.

By HARRIS & BROTHER, Gas Meter Manufacturers, 1115 and 1117 Cherry Street, Philadelphia. Full bound in pocket-book form $2.00

Hats and Felting:

A Practical Treatise on their Manufacture. By a Practical Hatter. Illustrated by Drawings of Machinery, etc. 8vo. $1.25

HOFMANN.—A Practical Treatise on the Manufacture of Paper in all its Branches.

By CARL HOFMANN. Late Superintendent of paper mills in Germany and the United States; recently manager of the Public Ledger Paper Mills, near Elkton, Md. Illustrated by 110 wood engravings, and five large folding plates. In one volume, 4to., cloth; 398 pages $15.00

HUGHES.—American Miller and Millwright's Assistant.

By WM. CARTER HUGHES. A new edition. In one vol., 12mo. $1.50

HURST.—A Hand-Book for Architectural Surveyors and others engaged in Building:

Containing Formulæ useful in Designing Builder's work, Table of Weights, of the materials used in Building, Memoranda connected with Builders' work, Mensuration, the Practice of Builders' Measurement, Contracts of Labor, Valuation of Property, Summary of the Practice in Dilapidation, etc., etc. By J. F. HURST, C. E. Second edition, pocket-book form, full bound $2.50

JERVIS.—Railway Property:

A Treatise on the Construction and Management of Railways; designed to afford useful knowledge, in the popular style, to the holders of this class of property; as well as Railway Managers, Officers, and Agents. By JOHN B. JERVIS, late Chief Engineer of the Hudson River Railroad, Croton Aqueduct, etc. In one vol., 12mo., cloth $2.00

JOHNSTON.—Instructions for the Analysis of Soils, Limestones, and Manures.

By J. F. W. JOHNSTON. 12mo. 38

KEENE.—A Hand-Book of Practical Gauging:

For the Use of Beginners, to which is added, A Chapter on Distillation, describing the process in operation at the Custom House for ascertaining the strength of wines. By JAMES B. KEENE, of H. M. Customs. 8vo. $1.25

KELLEY.—Speeches, Addresses, and Letters on Industrial and Financial Questions.

By Hon. WILLIAM D. KELLEY, M. C. In one volume, 544 pages, 8vo. $3.00

KENTISH.—A Treatise on a Box of Instruments,

And the Slide Rule; with the Theory of Trigonometry and Logarithms, including Practical Geometry, Surveying, Measuring of Timber, Cask and Malt Gauging, Heights, and Distances. By THOMAS KENTISH. In one volume. 12mo. $1.25

KOBELL.—ERNI.—Mineralogy Simplified:

A short Method of Determining and Classifying Minerals, by means of simple Chemical Experiments in the Wet Way. Translated from the last German Edition of F. VON KOBELL, with an Introduction to Blow-pipe Analysis and other additions. By HENRI ERNI, M. D., late Chief Chemist, Department of Agriculture, author of "Coal Oil and Petroleum." In one volume, 12mo. $2.50

LANDRIN.—A Treatise on Steel:

Comprising its Theory, Metallurgy, Properties, Practical Working, and Use. By M. H. C. LANDRIN, Jr., Civil Engineer. Translated from the French, with Notes, by A. A. FESQUET, Chemist and Engineer. With an Appendix on the Bessemer and the Martin Processes for Manufacturing Steel, from the Report of Abram S. Hewitt, United States Commissioner to the Universal Exposition, Paris, 1867. In one volume, 12mo. $3.00

LARKIN.—The Practical Brass and Iron Founder's Guide:

A Concise Treatise on Brass Founding, Moulding, the Metals and their Alloys, etc.: to which are added Recent Improvements in the Manufacture of Iron, Steel by the Bessemer Process, etc., etc. By JAMES LARKIN, late Conductor of the Brass Foundry Department in Reany, Neafie & Co's. Penn Works, Philadelphia. Fifth edition, revised, with Extensive additions. In one volume, 12mo.   .   . $2.25

LEAVITT.—Facts about Peat as an Article of Fuel:

With Remarks upon its Origin and Composition, the Localities in which it is found, the Methods of Preparation and Manufacture, and the various Uses to which it is applicable; together with many other matters of Practical and Scientific Interest. To which is added a chapter on the Utilization of Coal Dust with Peat for the Production of an Excellent Fuel at Moderate Cost, specially adapted for Steam Service. By T. H. LEAVITT. Third edition. 12mo.   .   .   . $1.75

LEROUX, C.—A Practical Treatise on the Manufacture of Worsteds and Carded Yarns:

Comprising Practical Mechanics, with Rules and Calculations applied to Spinning; Sorting, Cleaning, and Scouring Wools; the English and French methods of Combing, Drawing, and Spinning Worsteds and Manufacturing Carded Yarns. Translated from the French of CHARLES LEROUX, Mechanical Engineer, and Superintendent of a Spinning Mill, by HORATIO PAINE, M. D., and A. A. FESQUET, Chemist and Engineer. Illustrated by 12 large Plates. To which is added an Appendix, containing extracts from the Reports of the International Jury, and of the Artisans selected by the Committee appointed by the Council of the Society of Arts, London, on Woollen and Worsted Machinery and Fabrics, as exhibited in the Paris Universal Exposition, 1867. 8vo., cloth. $5.00

LESLIE (Miss).—Complete Cookery:

Directions for Cookery in its Various Branches. By MISS LESLIE. 60th thousand. Thoroughly revised, with the addition of New Receipts. In one volume, 12mo., cloth. $1.50

LESLIE (Miss).—Ladies' House Book:

A Manual of Domestic Economy. 20th revised edition. 12mo., cloth.

LESLIE (Miss).—Two Hundred Receipts in French Cookery.

Cloth, 12mo.

LIEBER.—Assayer's Guide:

Or, Practical Directions to Assayers, Miners, and Smelters, for the Tests and Assays, by Heat and by Wet Processes, for the Ores of all the principal Metals, of Gold and Silver Coins and Alloys, and of Coal, etc. By OSCAR M. LIEBER. 12mo., cloth. . . $1.25

LOTH.—The Practical Stair Builder:

A Complete Treatise on the Art of Building Stairs and Hand-Rails, Designed for Carpenters, Builders, and Stair-Builders. Illustrated with Thirty Original Plates. By C. EDWARD LOTH, Professional Stair-Builder. One large 4to. volume. $10.00

LOVE.—The Art of Dyeing, Cleaning, Scouring, and Finishing, on the Most Approved English and French Methods:

Being Practical Instructions in Dyeing Silks, Woollens, and Cottons, Feathers, Chips, Straw, etc. Scouring and Cleaning Bed and Window Curtains, Carpets, Rugs, etc. French and English Cleaning, any Color or Fabric of Silk, Satin, or Damask. By THOMAS LOVE, a Working Dyer and Scourer. Second American Edition, to which are added General Instructions for the Use of Aniline Colors. In one volume, 8vo., 343 pages. $5.00

MAIN and **BROWN.**—Questions on Subjects Connected with the Marine Steam-Engine:

And Examination Papers: with Hints for their Solution. By THOMAS J. MAIN, Professor of Mathematics, Royal Naval College, and THOMAS BROWN, Chief Engineer, R. N. 12mo., cloth.   .   .   . $1.50

MAIN and **BROWN.**—The Indicator and Dynamometer:

With their Practical Applications to the Steam-Engine. By THOMAS J. MAIN, M. A. F. R., Assistant Professor Royal Naval College, Portsmouth, and THOMAS BROWN, Assoc. Inst. C. E., Chief Engineer, R. N., attached to the Royal Naval College. Illustrated. From the Fourth London Edition. 8vo. $1.50

MAIN and **BROWN.**—The Marine Steam-Engine.

By THOMAS J. MAIN, F. R.; Assistant S. Mathematical Professor at the Royal Naval College, Portsmouth, and THOMAS BROWN, Assoc. Inst. C. E., Chief Engineer R. N. Attached to the Royal Naval College. Authors of "Questions connected with the Marine Steam-Engine," and the "Indicator and Dynamometer." With numerous Illustrations. In one volume, 8vo. $5.00

MARTIN.—Screw-Cutting Tables, for the Use of Mechanical Engineers:

Showing the Proper Arrangement of Wheels for Cutting the Threads of Screws of any required Pitch; with a Table for Making the Universal Gas-Pipe Thread and Taps. By W. A. MARTIN, Engineer. 8vo.   50

Mechanics' (Amateur) Workshop:

A treatise containing plain and concise directions for the manipulation of Wood and Metals, including Casting, Forging, Brazing, Soldering, and Carpentry. By the author of the "Lathe and its Uses." Third edition. Illustrated. 8vo. $3.00

MOLESWORTH.—Pocket-Book of Useful Formulæ and Memoranda for Civil and Mechanical Engineers.

By GUILFORD L. MOLESWORTH, Member of the Institution of Civil Engineers, Chief Resident Engineer of the Ceylon Railway. Second American, from the Tenth London Edition. In one volume, full bound in pocket-book form. $2.00

NAPIER.—A System of Chemistry Applied to Dyeing.

By JAMES NAPIER, F. C. S. A New and Thoroughly Revised Edition. Completely brought up to the present state of the Science, including the Chemistry of Coal Tar Colors, by A. A. FESQUET, Chemist and Engineer. With an Appendix on Dyeing and Calico Printing, as shown at the Universal Exposition, Paris, 1867. Illustrated. In one volume, 8vo., 422 pages. $5.00

NAPIER.—Manual of Electro-Metallurgy:

Including the Application of the Art to Manufacturing Processes. By JAMES NAPIER. Fourth American, from the Fourth London edition, revised and enlarged. Illustrated by engravings. In one vol., 8vo. $2.00

NASON.—Table of Reactions for Qualitative Chemical Analysis.

By HENRY B. NASON, Professor of Chemistry in the Rensselaer Polytechnic Institute, Troy, New York. Illustrated by Colors. . 63

NEWBERY.—Gleanings from Ornamental Art of every style:

Drawn from Examples in the British, South Kensington, Indian, Crystal Palace, and other Museums, the Exhibitions of 1851 and 1862, and the best English and Foreign works. In a series of one hundred exquisitely drawn Plates, containing many hundred examples. By ROBERT NEWBERY. 4to. $15.00

NICHOLSON.—A Manual of the Art of Bookbinding:

Containing full instructions in the different Branches of Forwarding, Gilding, and Finishing. Also, the Art of Marbling Book-edges and Paper. By JAMES B. NICHOLSON. Illustrated. 12mo., cloth. $2.25

NICHOLSON.—The Carpenter's New Guide:

A Complete Book of Lines for Carpenters and Joiners. By PETER NICHOLSON. The whole carefully and thoroughly revised by H. K. DAVIS, and containing numerous new and improved and original Designs for Roofs, Domes, etc. By SAMUEL SLOAN, Architect. Illustrated by 80 plates. 4to. $4.50

NORRIS.—A Hand-book for Locomotive Engineers and Machinists:

Comprising the Proportions and Calculations for Constructing Locomotives; Manner of Setting Valves; Tables of Squares, Cubes, Areas, etc., etc. By SEPTIMUS NORRIS, Civil and Mechanical Engineer. New edition. Illustrated. 12mo., cloth. $2.00

NYSTROM.—On Technological Education, and the Construction of Ships and Screw Propellers:

For Naval and Marine Engineers. By JOHN W. NYSTROM, late Acting Chief Engineer, U. S. N. Second edition, revised with additional matter. Illustrated by seven engravings. 12mo. . . $1.50

O'NEILL.—A Dictionary of Dyeing and Calico Printing:

Containing a brief account of all the Substances and Processes in use in the Art of Dyeing and Printing Textile Fabrics; with Practical Receipts and Scientific Information. By CHARLES O'NEILL, Analytical Chemist; Fellow of the Chemical Society of London; Member of the Literary and Philosophical Society of Manchester; Author of "Chemistry of Calico Printing and Dyeing." To which is added an Essay on Coal Tar Colors and their application to Dyeing and Calico Printing. By A. A. FESQUET, Chemist and Engineer. With an Appendix on Dyeing and Calico Printing, as shown at the Universal Exposition, Paris, 1867. In one volume, 8vo., 491 pages. . $6.00

ORTON.—Underground Treasures:

How and Where to Find Them. A Key for the Ready Determination of all the Useful Minerals within the United States. By JAMES ORTON, A. M. Illustrated, 12mo. $1.50

OSBORN.—American Mines and Mining:

Theoretically and Practically Considered. By **Prof.** H. S. OSBORN. Illustrated by numerous engravings. **8vo.** (*In preparation.*)

OSBORN.—The Metallurgy of Iron and Steel:

Theoretical and Practical in all its Branches; with special reference to American Materials and Processes. By H. S. OSBORN, LL. D., Professor of Mining and Metallurgy in Lafayette College, Easton, Pennsylvania. Illustrated by numerous large folding plates and wood-engravings. 8vo. $15.00

OVERMAN.—The Manufacture of Steel:

Containing the Practice and Principles of Working and Making Steel. A Handbook for Blacksmiths and Workers in Steel and Iron, Wagon Makers, **Die** Sinkers, Cutlers, and Manufacturers of Files and Hard-**ware**, of Steel and Iron, and for Men of Science and Art. By FREDERICK OVERMAN, Mining Engineer, Author of the "Manufacture of Iron," etc. A new, enlarged, and revised Edition. By A. A. FESQUET, Chemist and Engineer. $1.50

OVERMAN.—The Moulder and Founder's Pocket Guide:

A Treatise on Moulding and Founding in Green-sand, Dry-sand, Loam, and Cement; the Moulding of Machine Frames, Mill-gear, Hollow-ware, Ornaments, Trinkets, Bells, and Statues; Description of Moulds for Iron, Bronze, Brass, and other Metals; Plaster of Paris, Sulphur, Wax, and other articles commonly used in Casting; the Construction of Melting Furnaces, the Melting and Founding of Metals; the Composition of Alloys and their Nature. With an Appendix containing Receipts for Alloys, Bronze, Varnishes and Colors for Castings; also, Tables on the Strength and other qualities of Cast Metals. By FREDERICK OVERMAN, Mining Engineer, Author of "The Manufacture of Iron." With 42 Illustrations. 12mo. $1.50

Painter, Gilder, and Varnisher's Companion:

Containing Rules and Regulations in everything relating to the Arts of Painting, Gilding, Varnishing, Glass-Staining, Graining, Marbling, Sign-Writing, Gilding on Glass, and Coach Painting and Varnishing; Tests for the Detection of Adulterations in Oils, Colors, etc.; and a Statement of the Diseases to which Painters are peculiarly liable, with the Simplest and Best Remedies. Sixteenth Edition. Revised, with an Appendix. Containing Colors and Coloring—Theoretical and Practical. Comprising descriptions of a great variety of Additional Pigments, their Qualities and Uses, to which are added, Dryers, and Modes and Operations of Painting, etc. Together with Chevreul's Principles of Harmony and Contrast of Colors. 12mo., cloth. $1.50

PALLETT.—The Miller's, Millwright's, and Engineer's Guide.

By HENRY PALLETT. Illustrated. In one volume, 12mo. $3.00

PERCY.—The Manufacture of Russian Sheet-Iron.

By JOHN PERCY, M.D., F.R.S., Lecturer on Metallurgy at the Royal School of Mines, and to The Advanced Class of Artillery Officers at the Royal Artillery Institution, Woolwich; Author of "Metallurgy." With Illustrations. 8vo., paper.   .   .   .   .   .   .   50 cts.

PERKINS.—Gas and Ventilation.

Practical Treatise on Gas and Ventilation. With Special Relation to Illuminating, Heating, and Cooking by Gas. Including Scientific Helps to Engineer-students and others. With Illustrated Diagrams. By E. E. PERKINS. 12mo., cloth. $1.25

PERKINS and STOWE.—A New Guide to the Sheet-iron and Boiler Plate Roller:

Containing a Series of Tables showing the Weight of Slabs and Piles to produce Boiler Plates, and of the Weight of Piles and the Sizes of Bars to produce Sheet-iron; the Thickness of the Bar Gauge in decimals; the Weight per foot, and the Thickness on the Bar or Wire Gauge of the fractional parts of an inch; the Weight per sheet, and the Thickness on the Wire Gauge of Sheet-iron of various dimensions to weigh 112 lbs. per bundle; and the conversion of Short Weight into Long Weight, and Long Weight into Short. Estimated and collected by G. H. PERKINS and J. G. STOWE. $2.50

PHILLIPS and DARLINGTON.—Records of Mining and Metallurgy;

Or Facts and Memoranda for the use of the Mine Agent and Smelter. By J. ARTHUR PHILLIPS, Mining Engineer, Graduate of the Imperial School of Mines, France, etc., and JOHN DARLINGTON. Illustrated by numerous engravings. In one volume, 12mo.   .   . $2.00

PROTEAUX.—Practical Guide for the Manufacture of Paper and Boards.

By A. PROTEAUX, Civil Engineer, and Graduate of the School of Arts and Manufactures, and Director of Thiers' Paper Mill, Puy-de-Dôme. With additions, by L. S. LE NORMAND. Translated from the French, with Notes, by HORATIO PAINE, A. B., M.D. To which is added a Chapter on the Manufacture of Paper from Wood in the United States, by HENRY T. BROWN, of the "American Artisan." Illustrated by six plates, containing Drawings of Raw Materials, Machinery, Plans of Paper-Mills, etc., etc. 8vo. $10.00

REGNAULT.—Elements of Chemistry.

By M. V. REGNAULT. Translated from the French by T. FORREST BETTON, M.D., and edited, with Notes, by JAMES C. BOOTH, Melter and Refiner U. S. Mint, and WM. L. FABER, Metallurgist and Mining Engineer. Illustrated by nearly 700 wood engravings. Comprising nearly 1500 pages. In two volumes, 8vo., cloth.   .   .   . $7.50

REID.—A Practical Treatise on the Manufacture of Portland Cement:

By HENRY REID, C. E. To which is added a Translation of M. A. Lipowitz's Work, describing a New Method adopted in Germany for Manufacturing that Cement, by W. F. REID. Illustrated by plates and wood engravings. 8vo. $6.00

RIFFAULT, VERGNAUD, and TOUSSAINT.—A Practical Treatise on the Manufacture of Varnishes.

By MM. RIFFAULT, VERGNAUD, and TOUSSAINT. Revised and Edited by M. F. MALEPEYRE and Dr. EMIL WINCKLER. Illustrated. In one volume, 8vo. (*In preparation.*)

RIFFAULT, VERGNAUD, and TOUSSAINT.—A Practical Treatise on the Manufacture of Colors for Painting:

Containing the best Formulæ and the Processes the Newest and in most General Use. By MM. RIFFAULT, VERGNAUD, and TOUSSAINT. Revised and Edited by M. F. MALEPEYRE and Dr. EMIL WINCKLER. Translated from the French by A. A. FESQUET, Chemist and Engineer. Illustrated by Engravings. In one volume, 650 pages, 8vo. $7.50

ROBINSON.—Explosions of Steam Boilers:

How they are Caused, and how they may be Prevented. By J. R. ROBINSON, Steam Engineer. 12mo. $1.25

ROPER.—A Catechism of High Pressure or Non-Condensing Steam-Engines:

Including the Modelling, Constructing, Running, and Management of Steam Engines and Steam Boilers. With Illustrations. By STEPHEN ROPER, Engineer. Full bound tucks . . . $2.00

ROSELEUR.—Galvanoplastic Manipulations:

A Practical Guide for the Gold and Silver Electro-plater and the Galvanoplastic Operator. Translated from the French of ALFRED ROSELEUR, Chemist, Professor of the Galvanoplastic Art, Manufacturer of Chemicals, Gold and Silver Electro-plater. By A. A. FESQUET, Chemist and Engineer. Illustrated by over 127 Engravings on wood. 8vo., 495 pages. $6.00

☞ This Treatise is the fullest and by far the best on this subject ever published in the United States.

SCHINZ.—Researches on the Action of the Blast Furnace.

By CHARLES SCHINZ. Translated from the German with the special permission of the Author by WILLIAM H. MAW and MORITZ MÜLLER. With an Appendix written by the Author expressly for this edition. Illustrated by seven plates, containing 28 figures. In one volume, 12mo. $4.25

SHAW.—Civil Architecture:

Being a Complete Theoretical and Practical System of Building, containing the Fundamental Principles of the Art. By EDWARD SHAW, Architect. To which is added a Treatise on Gothic Architecture, etc. By THOMAS W. SILLOWAY and GEORGE M. HARDING, Architects. The whole illustrated by One Hundred and Two quarto plates finely engraved on copper. Eleventh Edition. 4to., cloth. . $10.00

SHUNK.—A Practical Treatise on Railway Curves and Location, for Young Engineers.

By WILLIAM F. SHUNK, Civil Engineer. 12mo. . . $2.00

SLOAN.—American Houses:

A variety of Original Designs for Rural Buildings. Illustrated by 26 colored Engravings, with Descriptive References. By SAMUEL SLOAN, Architect, author of the "Model Architect," etc., etc. 8vo. $2.50

SMEATON.—Builder's Pocket Companion:

Containing the Elements of Building, Surveying, and Architecture; with Practical Rules and Instructions connected with the subject. By A. C. SMEATON, Civil Engineer, etc. In one volume, 12mo. $1.50

SMITH.—A Manual of Political Economy.

By E. PESHINE SMITH. A new Edition, to which is added a full Index. 12mo., cloth. $1.25

SMITH.—Parks and Pleasure Grounds:

Or Practical Notes on Country Residences, Villas, Public Parks, and Gardens. By CHARLES H. J. SMITH, Landscape Gardener and Garden Architect, etc., etc. 12mo. $2.25

SMITH.—The Dyer's Instructor:

Comprising Practical Instructions in the Art of Dyeing Silk, Cotton, Wool, and Worsted, and Woollen Goods: containing nearly 800 Receipts. To which is added a Treatise on the Art of Padding; and the Printing of Silk Warps, Skeins, and Handkerchiefs, and the various Mordants and Colors for the different styles of such work. By DAVID SMITH, Pattern Dyer. 12mo., cloth. . . . $3.00

SMITH.—The Practical Dyer's Guide:

Comprising Practical Instructions in the Dyeing of Shot Cobourgs, Silk Striped Orleans, Colored Orleans from Black Warps, Ditto from White Warps, Colored Cobourgs from White Warps, Merinos, Yarns, Woollen Cloths, etc. Containing nearly 300 Receipts, to most of which a Dyed Pattern is annexed. Also, A Treatise on the Art of Padding. By DAVID SMITH. In one volume, 8vo. Price. . . $25.00

STEWART.—The American System.

Speeches on the Tariff Question, and on Internal Improvements, principally delivered in the House of Representatives of the United States. By ANDREW STEWART, late M. C. from Pennsylvania. With a Portrait, and a Biographical Sketch. In one volume, 8vo., 407 pages. $3.00

STOKES.—Cabinet-maker's and Upholsterer's Companion:

Comprising the Rudiments and Principles of Cabinet-making and Upholstery, with Familiar Instructions, illustrated by Examples for attaining a Proficiency in the Art of Drawing, as applicable to Cabinet-work; the Processes of Veneering, Inlaying, and Buhl-work; the Art of Dyeing and Staining Wood, Bone, Tortoise Shell, etc. Directions for Lackering, Japanning, and Varnishing; to make French Polish; to prepare the Best Glues, Cements, and Compositions, and a number of Receipts particularly useful for workmen generally. By J. STOKES. In one volume, 12mo. With Illustrations. . $1.25

Strength and other Properties of Metals:

Reports of Experiments on the Strength and other Properties of Metals for Cannon. With a Description of the Machines for testing Metals, and of the Classification of Cannon in service. By Officers of the Ordnance Department U. S. Army. By authority of the Secretary of War. Illustrated by 25 large steel plates. In one volume, 4to. . $10.00

SULLIVAN.—Protection to Native Industry.

By Sir EDWARD SULLIVAN, Baronet, author of "Ten Chapters on Social Reforms." In one volume, 8vo. $1.50

Tables Showing the Weight of Round, Square, and Flat Bar Iron, Steel, etc.,

By Measurement. Cloth. 63

TAYLOR.—Statistics of Coal:

Including Mineral Bituminous Substances employed in Arts and Manufactures; with their Geographical, Geological, and Commercial Distribution and Amount of Production and Consumption on the American Continent. With Incidental Statistics of the Iron Manufacture. By R. C. TAYLOR. Second edition, revised by S. S. HALDEMAN. Illustrated by five Maps and many wood engravings. 8vo., cloth. $10.00

TEMPLETON.—The Practical Examinator on Steam and the Steam-Engine:

With Instructive References relative thereto, arranged for the Use of Engineers, Students, and others. By WM. TEMPLETON, Engineer. 12mo. $1.25

THOMAS.—The Modern Practice of Photography.

By R. W. THOMAS, F. C. S. 8vo., cloth. 75

THOMSON.—Freight Charges Calculator.

By ANDREW THOMSON, Freight Agent. 24mo. . . . $1.25

TURNING: Specimens of Fancy Turning Executed on the Hand or Foot Lathe:

With Geometric, Oval, and Eccentric Chucks, and Elliptical Cutting Frame. By an Amateur. Illustrated by 30 exquisite Photographs. 4to. $3.00

Turner's (The) Companion:

Containing Instructions in Concentric, Elliptic, and Eccentric Turning: also various Plates of Chucks, Tools, and Instruments; and Directions for using the Eccentric Cutter, Drill, Vertical Cutter, and Circular Rest; with Patterns and Instructions for working them. A new edition in one volume, 12mo. **$1.50**

URBIN.—BRULL.—A Practical Guide for Puddling Iron and Steel.

By ED. URBIN, Engineer of Arts and Manufactures. A Prize Essay read before the Association of Engineers, Graduate of the School of Mines, of Liege, Belgium, at the Meeting of 1865–6. To which is added A COMPARISON OF THE RESISTING PROPERTIES OF IRON AND STEEL. By A. BRULL. Translated from the French by A. A. FESQUET, Chemist and Engineer. In one volume, 8vo. **$1.00**

VAILE.—Galvanized Iron Cornice-Worker's Manual:

Containing Instructions in Laying out the Different Mitres, and Making Patterns for all kinds of Plain and Circular Work. Also, Tables of Weights, Areas and Circumferences of Circles, and other Matter calculated to Benefit the Trade. By CHARLES A. VAILE, Superintendent "Richmond Cornice Works," Richmond, Indiana. Illustrated by 21 Plates. In one volume, 4to. **$5.00**

VILLE.—The School of Chemical Manures:

Or, Elementary Principles in the Use of Fertilizing Agents. From the French of M. GEORGE VILLE, by A. A. FESQUET, Chemist and Engineer. With Illustrations. In one volume, 12 mo. . . **$1.25**

VOGDES.—The Architect's and Builder's Pocket Companion and Price Book:

Consisting of a Short but Comprehensive Epitome of Decimals, Duodecimals, Geometry and Mensuration; with Tables of U. S. Measures, Sizes, Weights, Strengths, etc., of Iron, Wood, Stone, and various other Materials, Quantities of Materials in Given Sizes, and Dimensions of Wood, Brick, and Stone; and a full and complete Bill of Prices for Carpenter's Work; also, Rules for Computing and Valuing Brick and Brick Work, Stone Work, Painting, Plastering, etc. By FRANK W. VOGDES, Architect. Illustrated. Full bound in pocket-book form. **$2.00**
Bound in cloth. **1.50**

WARN.—The Sheet-Metal Worker's Instructor:

For Zinc, Sheet-Iron, Copper, and Tin-Plate Workers, etc. Containing a selection of Geometrical Problems; also, Practical and Simple Rules for describing the various Patterns required in the different branches of the above Trades. By REUBEN H. WARN, Practical Tin-plate Worker. To which is added an Appendix, containing Instructions for Boiler Making, Mensuration of Surfaces and Solids, Rules for Calculating the Weights of different Figures of Iron and Steel, Tables of the Weights of Iron, Steel, etc. Illustrated by 32 Plates and 37 Wood Engravings. 8vo. **$3.00**

WARNER.—New Theorems, Tables, and Diagrams for the Computation of Earth-Work:

Designed for the use of Engineers in Preliminary and Final Estimates, of Students in Engineering, and of Contractors and other non-professional Computers. In Two Parts, with an Appendix. Part I.—A Practical Treatise; Part II.—A Theoretical Treatise; and the Appendix. Containing Notes to the Rules and Examples of Part I.; Explanations of the Construction of Scales, Tables, and Diagrams, and a Treatise upon Equivalent Square Bases and Equivalent Level Heights. The whole illustrated by numerous original Engravings, comprising Explanatory Cuts for Definitions and Problems, Stereometric Scales and Diagrams, and a Series of Lithographic Drawings from Models, showing all the Combinations of Solid Forms which occur in Railroad Excavations and Embankments. By JOHN WARNER, A. M., Mining and Mechanical Engineer. 8vo. $5.00

WATSON.—A Manual of the Hand-Lathe:

Comprising Concise Directions for working Metals of all kinds, Ivory, Bone and Precious Woods; Dyeing, Coloring, and French Polishing; Inlaying by Veneers, and various methods practised to produce Elaborate work with Dispatch, and at Small Expense. By EGBERT P. WATSON, late of "The Scientific American," Author of "The Modern Practice of American Machinists and Engineers." Illustrated by 78 Engravings. $1.50

WATSON.—The Modern Practice of American Machinists and Engineers:

Including the Construction, Application, and Use of Drills, Lathe Tools, Cutters for Boring Cylinders, and Hollow Work Generally, with the most Economical Speed for the same; the Results verified by Actual Practice at the Lathe, the Vice, and on the Floor. Together with Workshop Management, Economy of Manufacture, the Steam-Engine, Boilers, Gears, Belting, etc., etc. By EGBERT P. WATSON, late of the "Scientific American." Illustrated by 86 Engravings. In one volume, 12mo. $2.50

WATSON.—The Theory and Practice of the Art of Weaving by Hand and Power:

With Calculations and Tables for the use of those connected with the Trade. By JOHN WATSON, Manufacturer and Practical Machine Maker. Illustrated by large Drawings of the best Power Looms. 8vo. $10.00

WEATHERLY.—Treatise on the Art of Boiling Sugar, Crystallizing, Lozenge-making, Comfits, Gum Goods.

12mo. $2.00

WEDDING.—The Metallurgy of Iron;

Theoretically and Practically Considered. By Dr. HERMANN WEDDING, Professor of the Metallurgy of Iron at the Royal Mining Academy, Berlin. Translated by JULIUS DU MONT, Bethlehem, Pa. Illustrated by 207 Engravings on Wood, and three Plates. In one volume, 8vo. (*In press.*)

WILL.—Tables for Qualitative Chemical Analysis.

By Professor HEINRICH WILL, of Giessen, Germany. Seventh edition. Translated by CHARLES F. HIMES, Ph. D., Professor of Natural Science, Dickinson College, Carlisle, Pa.   .   .   . $1.50

WILLIAMS.—On Heat and Steam:

Embracing New Views of Vaporization, Condensation, and Explosions. By CHARLES WYE WILLIAMS, A. I. C. E. Illustrated. 8vo. $3.50

WOHLER.—A Hand-Book of Mineral Analysis.

By F. WOHLER, Professor of Chemistry in the University of Göttingen. Edited by HENRY B. NASON, Professor of Chemistry in the Rensselaer Polytechnic Institute, Troy, New York. Illustrated. In one volume, 12mo. $3 00

WORSSAM.—On Mechanical Saws:

From the Transactions of the Society of Engineers, 1869. By S. W. WORSSAM, Jr. Illustrated by 18 large plates. 8vo.   .   . $5.00

www.ingramcontent.com/pod-product-compliance
Lightning Source LLC
Chambersburg PA
CBHW021709210326
41599CB00013B/1581